To Lynette . . .
Congratulations on winning
this giveaway book! I hope
you enjoy it.

THE COMPLETE STORY OF THE
WORLDWIDE INVASION OF THE
ORANGE ORBS

After reading the book, I
think you will agree that
truth is, indeed, stranger
than fiction. If you would
be so kind, I hope you will
write a review of my book.
Thanks!

TERRY RAY

Personal comments & questions :

tray@auxmail.IUP.edu

all the best —

SUNBURY PRESS

Mechanicsburg, Pennsylvania   USA

MW00809292

Published by Sunbury Press, Inc.
50 West Main Street, Suite A
Mechanicsburg, Pennsylvania 17055

www.sunburypress.com

For information about special discounts for bulk purchases, please contact Sunbury Press Orders Dept. at (855) 338-8359 or orders@sunburypress.com.

To request one of our authors for speaking engagements or book signings, please contact Sunbury Press Publicity Dept. at publicity@sunburypress.com.

ISBN: 978-1-62006-447-4 (Trade Paperback)
ISBN: 978-1-62006-448-1 (Mobipocket)
ISBN: 978-1-62006-449-8 (EPub)

FIRST SUNBURY PRESS EDITION: September 2014

Product of the United States of America
0 1 1 2 3 5 8 13 21 34 55

Set in Bookman Old Style
Designed by Lawrence Knorr
Cover by Lawrence Knorr
Edited by Lsawrence Knorr

Continue the Enlightenment!

ABOUT THE AUTHOR

Terry Ray is a former military pilot and trial lawyer, a retired law professor emeritus, and a certified Field Investigator for the Mutual Unidentified Flying Object Network, MUFON. He is currently a novelist for Sunbury Press and his recent novel, **GXM 731,** is a story of a voyage by beings from a distant galaxy to Earth told from the cosmic visitors' point of view. This book is Terry's first nonfiction work. He was inspired to write it after his own experience with orange orbs in July of 2013. This may be Terry's one and only nonfiction book after discovering how much harder it is to write nonfiction than novels.

Says Terry, "In a novel, the facts don't matter very much because you just make them up as you go. With nonfiction, you've got to go search the available data, determine what is relevant, and make sure it's accurate, then try to put it altogether into an organized and readable story—way too much work."

Terry is very thankful to all who assisted in the effort to get this book into print. The following are some of those who, without their help and support, this book would not exist:

Lawrence Knorr, owner of Sunbury Press, for having the courage to publish a nonfiction book about UFOs. Many mainstream publishers, Terry says, won't touch a nonfiction UFO book for fear of looking like a "fringe" publisher. Lawrence was convinced of the integrity of Terry's research and writing and, to his credit, accepted it for publishing.

MUFON members Jan Harzan, John Ventre, Fred Sluga, Sam Colosimo, and all the many Field Investigators, past and present. Jan, MUFON President, for his interest and encouragement in this book project; John, PA MUFON State Coordinator, Fred, PA Section Coordinator, and Sam, PA Field Investigator, for being my buddies and teaching me the ropes of UFO investigation; all of the Field Investigators who spent countless hours faithfully carrying out their UFO sighting assignments and building the massive MUFON database of cases which served as the essential tool for all of the information found in this book.

Bill Birnes, Mr. UFO, with whom Terry spent many hours discussing the writing of this book and who gave him many suggestions and invaluable insight.

Phyllis Webb, who was tireless in making this book into an organized, readable, and well-edited work of nonfiction.

Terry expresses his greatest thanks to his family, including his nine children, who are a never-ending source of support and inspiration for his writing.

CONTENTS

PART ONE

THE FACTS ABOUT
ORANGE ORBS

CHAPTER ONE
THE DISCOVERY

July 29, 2013

MUFON Case Sighting

The witness was on vacation with his family (wife, 3 children - ages 10, 11, 13) in Ocean City, MD, and was sitting on the ocean-front balcony of his first-floor hotel suite, #102, on the night of July 29, 2013, facing due east toward the Atlantic Ocean. (Quality Inn & Suites, 3301 Atlantic Ave., Ocean City, MD 21842, Phone: 410-289-1234, FAX: 410-289-0123) His wife and children were asleep inside the suite, and the lights were out, except for a nightlight. The witness' position was: N 38* 21.697' W 75* 04.263' Elevation: 20'. The night was clear with a minor degree of light diffusion from nearby hotels. The moon had not yet risen and would not for several hours thereafter. Beyond the horizon, due east, were occasional flashes of light emitted by a thunderstorm occurring beyond the Earth's curvature. Several meteor streaks were observed by the witness, as well as a number of commercial aircraft at standard cruising levels before the sighting.

At approximately 10:00 - 10:15 PM, the witness noticed a small round orange light to the north that appeared to be quite low over the ocean surface at approximately 100 – 300 feet in altitude, near the shore and above the ocean surface, moving on a southerly course toward the witness at a rather rapid rate of speed. Within approximately 10 seconds, the light had reached the size of a full moon and was very bright—the brightness and color approximated the coloring and intensity of a setting sun. The witness had binoculars and focused them on the light. With magnification, the object appeared to be a glowing orange sphere with uniform color and brightness. The color was soft and muted and emanated from the interior of the sphere. The sphere was pleasant and soothing to the eye. There were no other lights or visible characteristics appearing on the sphere. It was entirely silent.

The sphere was approximately five to ten miles north of the witness when it reached its maximum size as it stopped several hundred feet above the ocean surface and remained stationary for a short period of time. The sphere then began moving in an easterly direction, straight out to sea, and it diminished in size as its distance from the witness increased. It continued on its easterly course for approximately ten to fifteen seconds until its size appeared to be that of a very large star. At this point, the sphere commenced a straight climb of approximately forty-five degrees on a southerly heading. The climb was rapid and it leveled off at approximately 10,000 feet. It then continued straight and level on its southerly heading. The angle of inclination of the sphere from the witness, at this point, was approximately thirty degrees above the horizon. The sphere continued on its level southerly course for close to ten seconds, then turned to a south-easterly heading in the general direction of the thunderstorm and continued on this course until it was no longer visible, due to the increased distance from the witness.

The witness continued to watch the sky for approximately 5 minutes thereafter. Then, assuming the sighting had come to an end, he entered the suite and proceeded to the kitchen where he filled a bowl with caramel popcorn and returned to the balcony. The witness was absent from the balcony for approximately 5 minutes in procuring his snack. He resumed watching the sky for any further activity. After about 4 minutes, the witness noticed a second sphere in the same azimuth as he had initially seen the first sphere. This second sphere was identical to the first sphere in every respect. It proceeded to undertake the same maneuvers as the first sphere at the same speed, angles, and altitudes. It proceeded to the final south-easterly course and, like the first sphere, was no longer visible after a short time due to the increased distance from the witness. Approximately ten minutes after the second sphere's disappearance, a third sphere, identical to the first two, appeared in the same azimuth as the first two spheres and repeated the same course maneuvers identical to the first two spheres.

This process continued until a total of eight identical spheres were observed, at which time the sighting came to an end. The elapsed time, from beginning to end of the sighting, was approximately 90 minutes. All the spheres were entirely silent throughout their maneuvers and their speed remained constant. All legs of the course were flown in straight flight patterns, with no irregular movements, and at a constant velocity. The only aspect of their maneuvering that was not normal to Earth aircraft was their angular course changes, whereas a normal aircraft makes a curved pattern to accomplish a change in direction because of its reliance upon air pressure differentials to do so. Although the flight, velocity and climb characteristics of the spheres were unremarkable, the witness, having served as a former military pilot, had never observed any aircraft of this appearance in the sky and had never witnessed any craft under precise directional control do so in total silence. These aspects of the sighting were, to the witness, totally anomalous to his aviation experience. Based upon his aviation experience, the witness ruled out the possibility that these objects were balloons due to their precise controlled maneuvering and their velocity and climb characteristics. He also ruled out the

3

possibility of blimps due to their speed and maneuvering characteristics and, also, their physical appearance.

The foregoing report was filed with MUFON by the author of this book. At the time, I was going through the required exams and training to become a MUFON Field Investigator. MUFON is an abbreviation for "Mutual Unidentified Flying Object Network." This private non-profit organization was formed in 1969 to undertake a scientific systematic study of the UFO phenomenon that was occurring throughout the world at the time and which continues to this day. MUFON was founded in the United States, but has grown into a worldwide organization with chapters in most countries.

United States MUFON has its national headquarters in Newport Beach, California, and each state in the country has a MUFON chapter that works with national MUFON in a collaborative effort to pursue the scientific study of the UFO phenomenon. I am a Field Investigator for Pennsylvania MUFON. A certified Field Investigator is a MUFON member who has engaged in the extensive study of the UFO field and its history, has passed a comprehensive exam pertaining to these fields, and has undergone field training with an experienced Field Investigator. After successfully completing the training and examination, a Field Investigator is certified by the national office and is then qualified to undertake the investigation of cases assigned by the national office to the state MUFON chapter in which the sighting occurred. The state MUFON chapter then makes the assignment to a particular investigator(s). Within 48 hours of a case filing, the assigned Field Investigator(s) makes contact with the person filing the UFO sighting report, then proceeds to investigate the sighting and, upon completion of the investigation, files a final report with the national office, which becomes a permanent part of the MUFON case data bank.

Every case filed with MUFON is entered into the MUFON case files repository and is available to Field Investigators to conduct research and to publish their findings for the advancement of the scientific study of the UFO phenomenon and/or the benefit of the reading public, such as the book you are currently reading.

Membership in MUFON is currently expanding at a very rapid rate, as is the number of certified Field Investigators. Not every MUFON member is a Field Investigator and most are general members who are entitled to all the benefits of membership, such as receiving regular newsletters, UFO Journals of Research, and attendance at local, state, and national MUFON conferences, as well as having a degree of computer access to the central UFO case files that is very informative, although not as extensive as that available to Field Investigators.

The reason for the recent and rapid growth of MUFON is a matter of speculation, but those most familiar with the UFO field attribute the growth to the expanding number of television programs and movies dedicated to UFO's and aliens, plus the recent astronomical discoveries of the ever-growing numbers of planets throughout the cosmos that are earth-like and are very likely populated by life-forms—many of which could have intelligent beings who may have technology far beyond the sentient of Earth. MUFON is also likely growing because of the popularity of its new History Channel series, *HANGAR ONE*, which, each week, presents actual and compelling cases taken from its extensive central case repository.

Upon returning to Pennsylvania following my Ocean City sighting, I was anxious to tell my story to the veteran Field Investigators in my Pennsylvania region, anticipating their excitement. Such was not to be. Instead of excitement, my account was met with chuckles and a condescending explanation that, apparently, I should have known without them telling me. With amusement on their faces, it was explained to me that these so-called orange orbs are merely candle-lit, Chinese lanterns that are launched into the air at such events as festivals and weddings. "Everybody knows that," they said. I was advised to spend my time on something that was truly unidentified, since these floating orange objects had a simple and mundane explanation.

Perhaps other investigators-in-training would have simply accepted these pronouncements from veteran FI's and foregone any further involvement with the orange orbs but, as an experienced military pilot, I knew that what I saw could not have been candle-lit flying lanterns drifting along on a pleasant evening breeze. It was something else that I could not explain.

This experience spurred me to begin what turned out to be a year-long quest to uncover the mystery of what I saw that night from my balcony in Ocean City, Maryland. The quest ultimately resulted in the discovery of a worldwide phenomenon that could turn out to be the most significant event in the history of mankind.

CHAPTER TWO
PUTTING IT INTO PERSPECTIVE

In researching UFO's, there are a couple of things that the reader must understand before proceeding. First, organized research into the UFO field is in its infancy, having come into being in 1969 with the advent of MUFON. The information available is, therefore, quite limited as compared to long-standing fields of scientific research. Secondly, the only credible extant UFO data collection organization in the world is MUFON. It is, however, unknown to a large portion of the general public. Because of this, UFO sightings are often not reported to MUFON. As a matter of fact, the vast majority of UFO witnesses do not report their sightings to any entity of data collection whatsoever. The two prevalent reasons for failure to report are lack of knowledge as to where one should file a sighting report and the fear of being labeled as a "kook" for claiming to have seen a UFO.

If a person actually does report a sighting, the most common contacts are with police, media, and the military (none of which collects or admits to collecting data on UFOs). The problem with these choices varies with each entity. Police frequently blow-off such reports as nuisance calls, not wishing to take time away from what they consider more important matters. Media tend to use such reports with tongue-in-cheek condescending humor as the "and finally ...", funny, kooky story at the end of the newscast, concluding with a camera shot of a chortling, eye-rolling panel of newscasters.

I am not aware of any comprehensive studies that have been done that attempt to determine just how many people who witness a UFO actually report it to an official agency and, also, of those who do report to an official agency, how many file their report with MUFON. I have talked to a number of seasoned experts in the UFO field about this question and many say that, if we're lucky, perhaps one person in a hundred who witnesses a UFO actually files a report with MUFON. Until such a study is done, we will have to proceed with the general understanding that the number of sightings reported to MUFON is but a small fraction of all the UFO sightings that take place in the world every day.

The military, as an agent of the federal government, is under a long-standing federal directive to be uncooperative in UFO research and to actively make a mockery of anything pertaining to the extraterrestrial phenomenon. Thus, when anyone reports a UFO sighting to the government, they will uniformly downplay the sighting as something mundane (and often ridiculous) that the witness clearly mistook for a UFO. They will also then place it into their "nonexistent" database.

The government's reasons for taking this position are a matter of broad discussion within MUFON, but the most likely explanation is that the government: 1) does not wish to appear helpless in matters that may have national security implications (which they are in regard to UFOs) and 2) believes that if they openly admit to the existence of UFOs, it will result in generalized public panic.

The possibility of this happening was made real by the public reaction to Orson Welles' 1938 radio dramatization of the H.G. Wells novel about the invasion of Earth by extraterrestrials, *War of the Worlds.* Many thought the broadcast was real and millions across the country truly panicked because of it. No doubt the U.S. government took quite an interest in this reaction and very likely factored it into their position on UFOs. Thank you Orson.

Because of the many reasons cited above, it is not possible to come up with any firm numbers of actual UFO sightings. The only things that can be established are the trends and characteristics of particular sightings. For my research into orange orbs, I relied exclusively on MUFON case files as the only true scientifically organized, screened, investigated, and comprehensive repository of UFO reports in the world. A reader should keep in mind, however, that the numbers set forth in this book of orange orbs represent only a small fraction of the actual number of orbs that are flying in Earth's skies.

The first thing I discovered in my year-long orange orb research project was that not every observer who witnesses an orange orb reports it to MUFON in the same way. They made a variety of different choices among the selections available in MUFON's reporting system. The system offers: white, grey/lead, black, gold/copper, pink/rose, red, red/orange, yellow/orange, green, green/white, blue-green, blue, blue-white, and violet as color choices. As for shape choices, there are: blimp, boomerang, bullet/missile, cigar, cone, chevron, circle, cross, square/rectangle, star-like, teardrop, triangle, fireball, diamond, disc, teardrop, sphere, or egg as the shape choices. There are no "orange" or "orb" choices.

In my preliminary review, using a variety of the available combinations of choices under which a witness could have reported to MUFON, I found that the categories of Red/Orange and Sphere were the most common choices and so I chose to use these for my research. Researching all the possible combinations of all shapes and colors would be a near impossibility. I realized that, by using only one shape and one color, I would miss a good number of other cases involving orange orb sightings, but at least I could develop a wide variety of reports (I found over 2,000 cases under this combination) and would be able to track trends in the orb sightings that would likely apply to all the other combinations.

As for the historical span of my research, I decided to begin with the beginning of MUFON in 1969 and end the search at December 31, 2013.

CHAPTER THREE
BY THE NUMBERS

In 1969 there were only two red/orange ("orange" for the purposes of this book) spheres ("orbs" for the purposes of this book) sighted in the United States. For the next 33 years, the annual sightings were also very low, with an average of about four per year, with a couple of anomalous years such as 9 orbs in 1997, 10 in 1999, and 11 in 2002. The total number of orbs sighted in 33 years—from 1969 through 2002—was 130.

This is not to say that there were no orange orb sightings before 1969. In reviewing the historical archives of the MUFON database, which go back to 1890, I found there were eleven orange orbs sighted between 1890 and 1969 in the United States. Interestingly, all eleven orb sightings were clustered between 1946 and 1969. One orb was sighted in the 1940s, three in the 1950s, and seven in the 1960s.

One may speculate on why this clustering of sightings occurred as it did. Having grown up in the sixties, I can attest to the fact that it was a decade filled with riots, burning cities, and the Vietnam War. Perhaps such widespread conflagration attracted the attention of our cosmic visitors. The 1946 sighting was close, in date, to the famous Roswell, New Mexico incident where the U.S. government announced it had found the remains of a recently crashed extraterrestrial vehicle and bodies of cosmic visitors who were aboard the craft. (They, of course, quickly retracted this story for reasons still not explained.) Also, the first atomic explosions took place in the 1940s, which our cosmic visitors would have found interesting, as well. The 1950s sightings corresponded to the years of the Korean War and a greatly increased number of atomic detonations.

These sightings and Earth events line up with one another, but whether or not they are connected in a causative way is clearly open to debate. Perhaps, one day, the visitors, themselves, may explain their thinking to us.

During the decade from 2003 through 2013, things changed dramatically in regard to orange orb sightings. During the last ten years there were 2,255 orbs sighted in American skies. The relative year-to-year increase was also significant. For example, in 2003 there were five orbs sighted and in 2004 the total was ten. In 2005 there were thirteen orbs sighted and the following year there were sixteen. In 2007, the count jumped to 46, followed by 78 in 2008. 102 orbs were sighted in 2009 and 164 in 2010. 2011 had 305 and 2012 came in with 648. Last year, in 2013, 868 orbs were sighted over the U.S.

Although this book is exclusively about orange orb sightings above the United States, the reader should be aware that this tremendous yearly increase in the number of orbs prowling our atmosphere is not just an American phenomenon. The orb invasion is going on worldwide.

To put the actual number of orange orbs sighted in the U.S. in 2013 into a meaningful perspective, the reader should recall the realities cited above. The number of orbs reported to MUFON represents only a small portion of the actual number of orbs flying in our American airspace and the total number of orange orbs reported to MUFON, under all possible categories, greatly exceeds the number I found in my particular category of research.

To explain, while MUFON recorded 868 orange orb sightings in 2013 in my category of research, the actual number of orbs traversing the U.S. in 2013 is a significantly higher figure. If the one-in-a-hundred rule of thumb is even close to being accurate, then the true number of orange orbs in American skies in 2013 is probably closer to 86,800 and the true number of orbs actually sighted over the past decade in the U.S. is probably nearer a quarter of a million. The true worldwide number of orange orbs would be an extraordinary figure.

The orange orbs cruising above the U.S. are not evenly distributed throughout the year. There are, for instance, consistently fewer orbs sighted in the first half of the year than in the second half. In 2011, for example, 70 were seen in the first half of the year while 235 were seen in the second half. In 2012, 267 appeared in the first, 381 in the second. In 2013 there were 230 in the first and 638 in the second.

The reason for this phenomenon is, of course, anybody's guess, but one logical and simple explanation is that the weather in the United States is generally more pleasant and accommodating for outdoor activity during the last six months of the year than the first six. With more outdoor activity comes a greater chance of noticing nighttime phenomenon.

There is another fascinating anomaly in the date of orange orb appearances. Every year there are two particular evenings when the number of orange orb sightings dramatically spike to the highest single-day sightings. I will let the reader ponder the dates of these two high spiking evenings. In the chapter entitled "What?", found in Part Three of this book, I reveal the answer. (And now we will pause while the reader jumps to the "What?" chapter to see if his answer is correct.)

On the other hand, the orbs are quite consistent as to the time of day they make their appearance. The critical mass of orange orbs are seen year-round, between 8:30 and 10:30 at night. There are a few anomalies in the sighting cycle—one being that the orbs will be seen frequently in one area for a certain time period, then few are seen for a time period in that area, then the observations are, again, frequent in the area, and so on. For some reason, the cosmic visitors change the deployment of their space vehicles on this cyclical basis. I don't have any plausible hypothesis as to why.

As to where, geographically, these orbs are seen in the United States, they are certainly not evenly distributed. Some states have had hundreds of orange orb sightings during the last decade while some have as few as one.

To provide the reader with a clearer idea of the flight patterns of the orange orbs over the U.S., you will find the following illustrations at the end of this chapter that provide a great amount of detail pertaining to this uneven distribution of orb flights:

1. A map of the U.S. with the all orb sightings from 2003 – 2013 marked in the county where the sighting took place.
2. A nighttime satellite photo of the U.S.
3. A map of the U.S. showing the four orb sectors and the locations of the underwater orb bases from which the orbs are deployed.
4. A chart showing the number of orange orb sightings for each state in the U.S., including the District of Columbia, for each year from 2003 through 2013 and the totals for the decade.

Explanations of the illustrations will be found in this chapter and in the chapters that follow.

Pertaining to the orb chart, the reader will notice that the deployment of the orb flights sometimes change dramatically from year to year. In Missouri, for instance, the sightings suddenly spike from 21 in 2012 to 88 in 2013 (the highest reported yearly number in the entire decade and the greatest one year increase as well). In Michigan, the number of sighted orbs drops precipitously from 57 in 2012 to 21 in 2013. California jumps from 30 in 2011 to 78 in 2012. Oregon jumps from 11 in 2012 to 50 in 2013. Colorado jumps from 12 in 2012 to 34 in 2013. Utah jumps from 10 in 2012 to 25 in 2013.

What would cause our cosmic visitors to make such dramatic changes in their traffic patterns over certain states? Frankly, I cannot even suggest a logical reason as to why they would choose to do this. I greatly doubt it is merely some flippant change of mind and would assume there is a rationale for making these changes, but we will probably have to wait until our visitors decide to tell us.

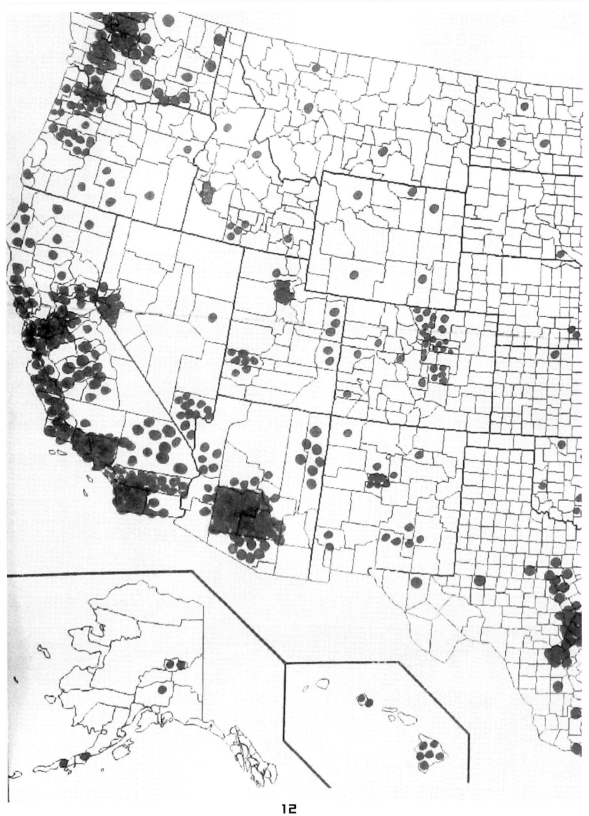

1. A map of the U.S. with the all orb sightings from 2003 – 2013 marked
   in the county where the sighting took place.

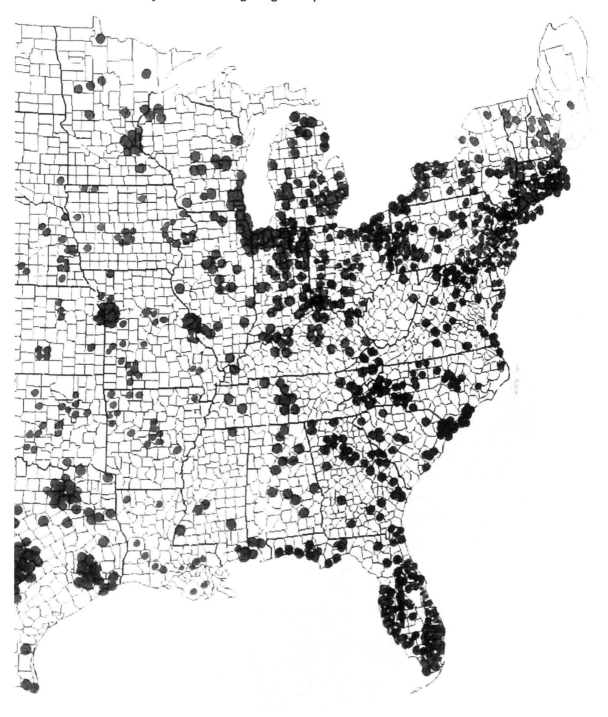

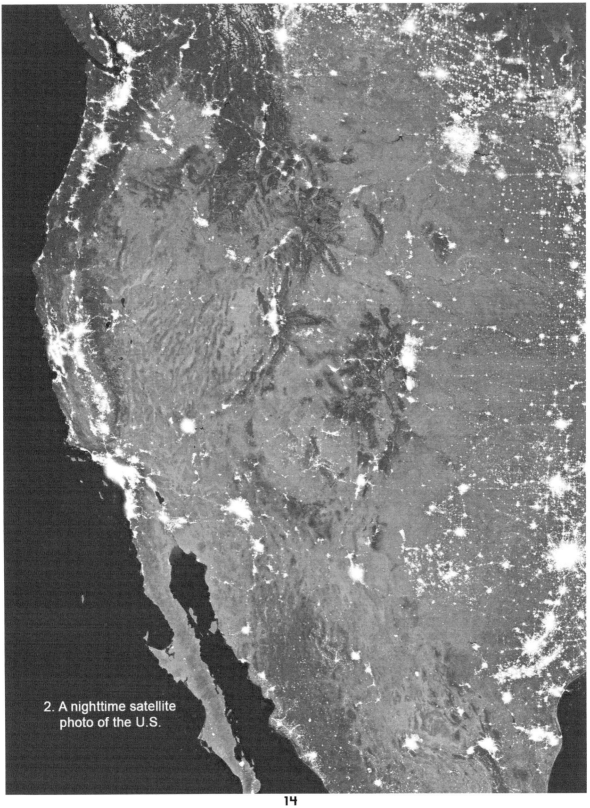

2. A nighttime satellite photo of the U.S.

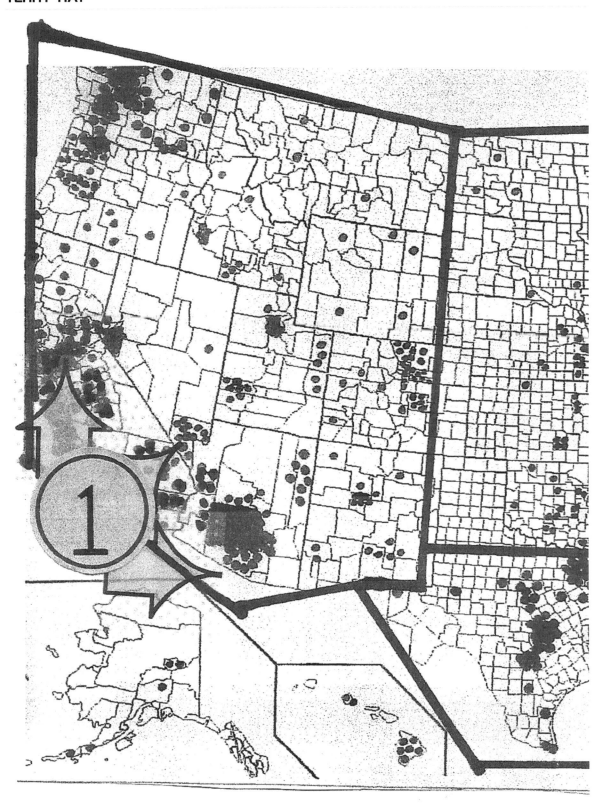

3. A map of the U.S. showing the four orb sectors and the locations of the underwater orb bases from which the orbs are deployed.

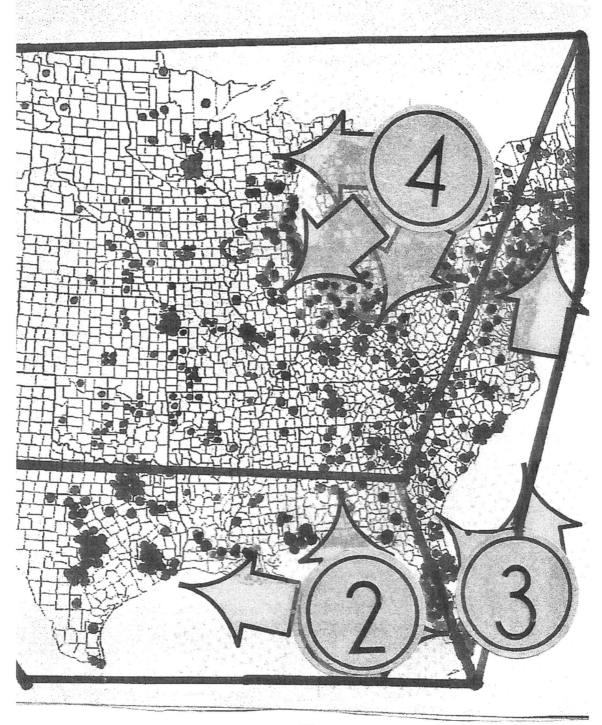

Orange Orb Sightings by Year

	State	2003	2004	2005	2006	2007	2008	2009	2010	2011	2012	2013	Total
1	California	2	3	1	3	5	12	14	18	30	78	86	252
2	Florida	0	0	0	1	3	2	2	9	20	49	55	141
3	Missouri	0	1	0	1	1	2	3	5	10	21	88	132
4	Michigan	0	0	0	1	2	6	7	9	18	57	21	121
5	Pennsylvania	0	0	1	0	1	3	3	11	19	38	41	117
6	Arizona	1	0	4	0	2	3	11	6	15	34	23	99
7	Ohio	0	0	0	1	2	2	8	3	14	26	38	94
8	Texas	0	0	0	0	1	12	9	7	6	32	14	81
9	New York	0	0	0	1	1	3	2	7	11	29	26	80
10	Oregon	0	1	1	0	2	1	2	3	8	11	50	79
11	Washington	0	0	0	2	2	2	3	6	4	23	28	70
12	Illinois	0	1	0	1	3	2	2	6	15	23	16	69
13	Colorado	0	1	0	0	1	4	2	4	4	12	34	62
14	Indiana	0	1	0	1	3	1	2	5	12	22	12	59
15	North Carolina	0	0	0	0	1	3	2	3	7	21	14	51
16	Iowa	0	0	0	0	1	0	0	2	5	11	28	47
17	Utah	0	0	0	0	0	1	0	1	3	10	25	40
18	Georgia	0	0	1	0	3	1	5	6	5	13	5	39
19	Arkansas	0	0	1	0	1	0	0	1	5	4	23	35
20	Nevada	0	0	1	0	0	0	2	0	4	7	21	35
21	Oklahoma	0	0	0	0	0	4	2	0	3	6	20	35
22	New Mexico	0	0	0	0	0	2	4	3	3	1	21	34
23	Idaho	0	0	0	0	0	1	1	2	7	8	15	34
24	Kansas	0	0	0	1	0	0	0	4	5	3	20	33
25	New Jersey	0	0	0	0	0	1	1	2	7	8	14	33
26	Massachusetts	0	0	0	0	1	0	0	4	6	13	9	33
27	South Carolina	1	0	0	0	2	0	0	2	7	12	8	32
28	Wisconsin	0	0	0	0	0	2	0	1	6	7	13	29
29	Virginia	0	0	0	0	2	1	0	2	4	9	11	29
30	Minnesota	0	0	0	1	1	1	1	0	3	9	12	28
31	Tennessee	0	0	0	0	0	1	2	8	1	6	6	24
32	Louisiana	0	0	0	0	0	1	1	1	2	3	15	23
33	Kentucky	0	0	0	1	0	0	3	2	6	5	6	23
34	Connecticut	0	0	0	0	1	1	1	3	2	5	6	19
35	Nebraska	0	0	1	0	1	0	3	5	7	1	0	18
36	Hawaii	1	0	0	0	0	0	0	2	1	1	9	14
37	Alabama	0	1	1	0	0	0	0	3	4	4	0	13
38	Wyoming	0	0	0	0	0	0	0	1	2	1	6	10
39	Alaska	0	1	0	0	0	0	0	1	2	1	5	10
40	New Hampshire	0	0	0	1	0	0	0	1	3	4	1	10
41	Maryland	0	0	0	0	1	0	2	1	2	7	7	9
42	North Dakota	0	0	0	0	2	0	0	0	0	1	4	7
43	South Dakota	0	0	0	0	0	3	0	0	0	0	3	6
44	Montana	0	0	0	0	0	0	1	1	0	0	3	5
45	Delaware	0	0	0	0	0	0	0	0	1	2	2	5
46	Vermont	0	0	0	0	0	0	0	1	1	1	2	5
47	Rhode Island	0	0	1	0	0	0	0	0	0	3	1	5
48	Maine	0	0	0	0	0	0	0	1	2	2	0	5
49	West Virginia	0	0	0	0	0	0	1	0	2	2	0	5
50	Mississippi	0	0	0	0	0	0	0	1	1	1	1	4
51	Dist. Of Col.	0	0	0	0	0	0	0	0	0	1	0	1
	Totals	5	10	13	16	46	78	102	164	305	648	868	2255

18

CHAPTER FOUR
TRAFFIC PATTERNS

Over a period of nearly a year, I reviewed all 2,255 cases filed with MUFON for the last decade of 2003 – 2013 under the classification of red/orange spheres. From this review, I was able to determine the county in which each of these orbs was sighted. After compiling this information, I created a map (prior page) with the outline of every county in the U.S., then placed a dot for every sighted orange orb in the appropriate county of sighting. You can see the final product on this map. As you will notice, the orb sightings are clustered into certain counties in the U.S. This raises the question as to why they would chose to visit certain areas very frequently and ignore other areas of the country. The answer may be found by looking at the illustration which follows the county map.

This illustration is a nighttime satellite photograph taken of the U.S. If you look at the streetlight pattern in the photo and compare it to the orb clusters on the map, the similarity is unmistakable. They match-up almost perfectly. It is obvious, then, that our cosmic visitors have chosen to fly their orange-cloaked space vehicles to where the lights are. The streetlight clusters are, of course, the lights of our major cities—and our major cities are our densest population hubs.

Our visitors, of course, know this. So in this regard, we can read the minds of our cosmic visitors. They have made the choice to fly their brightly-colored orbs to where the greatest number of people will see them. Why they want to be seen by the maximum number of Earthly beings on a regular basis, every night, will be addressed in the "Why" chapter in this book.

While our visitors are flying over our most populated areas in the country, there is one glaring anomaly which can be seen on the map and even better observed on the orb chart. Every large city in the U.S. is targeted for frequent nighttime orb visitations with only one exception: our nation's capital. In the entire history of orb documentation, there has been only one sighting of an orb over Washington, D.C. Why in the world would our cosmic visitors choose to almost totally avoid flying over such a large city? I can come up with only one possible explanation.

This center of our nation's government was attacked by terrorists on September 11, 2001. A plane was flown into the Pentagon and a second plane was headed toward either the White House or the Capitol Building before it was brought down in a Pennsylvania field by courageous Americans in our nation's first homeland battle with the terrorists. This attack, which was meant to decapitate the leadership of the United States, was a wake-up call for America. It clearly demonstrated the stark vulnerability of our nation's capital to a disastrous attack from our enemies. With this realization, we went about making Washington, D.C. the most heavily fortified airspace in the country with the placement of a mighty arsenal of surface-to-air missiles, throughout the city, to instantly destroy any hostile intruder. And, at the nearby military airfields, fighter jets sit on runways armed and ready to respond, at a moment's notice, to any intrusion.

Our cosmic visitors, who are capable of technology beyond our comprehension and who can travel many light years from their home to ours, are obviously very intelligent. Given this, they must certainly have the capability to detect that our nation's capital is very heavily fortified and that a violation of this airspace would, almost certainly, provoke an enormous and spectacular armed response.

Knowing this, they have clearly decided to avoid such a confrontation with our military, at the present time at least, by totally bypassing our capital's airspace. This reveals at least a part of our visitors' psyche. They are not, apparently (at least not at this juncture), looking for confrontation with us. This may be simply convenient for their present purposes, or it could mean that they truly come in peace to our home. We just don't know, at this moment, but will likely find out what's behind this decision to avoid our nation's capital by the nature of their future activities.

If the reader studies the orb map, it clearly appears that there are distinct sectors of orb concentration: Pacific, Northeast/Central, Atlantic, and Gulf. In reading each case, I was able to chart the time of night for each orb sighting and the direction in which it was traveling at that time. A clear pattern emerged from this charting. Orbs in each of the four sectors showed a consistent pattern of to-and-fro movement in relationship to the time of night they were flying. They deployed from their base of operations at about 8:30 P.M. and returned at about 10:30 P.M.

In the Pacific Sector, the orbs appeared to emanate from an area in the Pacific Ocean just off the coast of Los Angeles, with most traveling north, along the coast, at approximately 8:30 P.M. Pacific time, as far north as the state of Washington and some travel to the east as far as the border of Texas. At approximately 10:30 P.M., these orbs were seen returning to their point of origin. Occasionally, there were a few slight anomalies to these patterns as to the time of their travel.

In the Atlantic Sector, the orbs appeared to emanate from an area just off the coast of Miami, Florida in the Atlantic Ocean. They traveled northward, along the Atlantic coast, as far north as Maine, then returned to their base.

In the Gulf Sector, the orbs deployed from a base just off the coast of Sarasota,

Florida, traveled along the west coast of Florida, and continued along the Gulf coast through Texas before returning home.

In the Eastern/Central Sector, the orbs appeared to emanate from a point just off Presque Isle in Lake Erie. They flew in a number of different directions—west, east, and south—to areas of the United States that the other sector orbs did not visit.

This begs th "parked" when they were not in the air. Given th area over water in all four sectors, it hey "parked" in underwater bases, ben is conclusion there was eye witness tes these locations. Since these orbs woul , placing them underwater would be a

The Se base locations, the sectors, and traffic he sectors.

PART TWO

EYE WITNESS REPORTS

The following chapters contain reports written by eyewitnesses who have actually seen orange orbs and/or witnessed other phenomenon associated with them. These reports are taken directly from the MUFON case files. The only editing that has been done to the witness reports is to correct punctuation, spelling, spacing, and paragraph layout to facilitate easier reading and comprehension. The actual stories have not been changed in any way. Also, any information that could reveal the personal identity of the witness or others mentioned in the report have been redacted. It is the long-standing policy of MUFON that the identity of persons filing reports with this network remain strictly confidential.

The witnesses filing reports of UFOs to MUFON are quite diverse, including a wide variety of races, ethnic groups, ages, genders, professions, and educational levels. Because of this, the reports vary in terms of writing skills. The reader should keep in mind that those who submit reports to MUFON are not auditioning as professional writers, but are simply ordinary individuals who have had an extraordinary experience and who feel a responsibility to share them with both experts and the general public. They come forward at risk to their jobs, family relationships, and public standing. This risk is due to the unfortunate bigotry against those who claim to have witnessed extraterrestrial phenomenon, brought about, primarily, by the federal government's well-documented strategy of mocking and marginalizing anyone who makes such claims. This government mindset has permeated all levels of American society, including the media.

Thus, the author asks the reader to overlook the differences among the witnesses' writing abilities and focus, instead, on what they are telling you about their experiences. All the witnesses who have taken the time and risk to file a report with MUFON deserve our heartfelt appreciation.

Each of the following chapters in Part Two of this book focuses on a different aspect of the orange orb phenomenon—from detailed descriptions of the orbs to close encounters with the cosmic visitors piloting them. By the time you have read this impressive collection of eyewitness reports, you will have acquired a deep and personal understanding of the orange orb phenomenon and an appreciation of the ways these sightings have affected the lives of the witnesses. These witness reports are, of course, only a small portion of the total number of the 2,000+ cases I researched for this book and were selected as being representative of the many facets of the orange orb phenomenon.

So, turn the page and plunge into this treasure trove of real-life experiences that will sometimes be quite informative, sometimes fascinating, sometimes jaw-dropping, and sometimes downright frightening.

CHAPTER ONE
WHAT DO ORBS LOOK LIKE?

Witness # 1

On 6/28/2014, at approximately 2145 hours, I saw a bright blue sphere northeast of my location while traveling eastbound on Sample Road in Margate FL. While I saw the object in/from Margate, the object's apparent location placed it above Coconut Creek [city], and therefore NE of me. The object, while apparently large, was at quite a distance, and I wouldn't dare to guess its size nor actual distance from me at the time. The object hovered in the sky, while pulsating. It had what appeared to be a smaller red dot/light to its left, which would appear and disappear on occasion. The red light did not appear to blink or pulsate, but rather would simply appear and disappear. This object would also seem to change into a triangular shape with its apex seemingly pointing in a W-NW direction.

Finally, after approximately 10 minutes the object changed course in a zig-zag pattern from where it emerged in a line - then one was just gone, then the other two disappeared. There were choppers and airplanes around, but not close to the orbs, and these were completely different - no sound, and moved crisply when they changed direction. They just disappeared.

I told my friend, those are UFOs, and he laughed at me but I knew this was something I hadn't seen before. They didn't belong there. I knew I was seeing something strange.

I was excited and kind of scared. I was a student pilot some years ago, learned to fly and land a plane. I know what planes are. These things just did not behave like a chopper or plane and anyway, they were bright orange and very fast in the way they changed direction, going from a triangular formation, hovered for maybe 40 seconds, formed a line then, boom - one was gone, then the other two. These were not conventional aircraft and they weren't fireworks. Fireworks don't hover, change direction, line up in formation, move slowly across the sky, and then disappear. Someone said maybe drones, but really? That big, bright, and doing maneuvers? What did I see?

Witness #2

I was talking to my girlfriend on the phone at around 10:20 PST when I saw a dull orange light. I thought it was a helicopter until I saw another light fall from it and illuminate the background. That's when I thought "Could be a UFO," so I ran and woke up my dad. In the video, after I handed the camera to him, he hits the "end" button but keeps holding it up like its filming so I assumed it was still rolling. It was not. Anyway, the light changed direction several times and dropped something ... it was luminous. This happened three times. It occasionally blinked out but for the most part it was dull orange. There were three witnesses to this: my parents and I. It moved off into the horizon where we lost sight of it. The whole event lasted around 5 minutes.

Witness #3

7: 25 EST 1/19/13 - in front of my house – I was sitting on the porch enjoying the weather, when I noticed a red orb flashing and then it blinked out and 3 appeared. It flew over my neighbor's house in a boomerang shape, went over the right side of my driveway, then over the trees. I heard my neighbor come outside, I ran to the side of my yard, yelled at him and he saw them. They were headed NE and then they turned and went West and a new orb joined them and it was flashing white and red. After that, heard a few military planes about 5 minutes later heading South over my neighbor's house. I did try to tape it on video, but my camera could not focus for some reason. I will submit the video when I look at it.

Witness #4

I was cleaning up my home office at midnight and noted when looking out the back window what a clear night it was. I then noticed what I first thought was a plane with blinking lights over Tennessee Valley, but it didn't move after a couple of minutes. It didn't appear like a helicopter although it was fairly stationary. When looking at it closely, it appeared like a blob of differently colored, sparkling lights. The sparkling lights were white, blue and light green.

After about 10 minutes it moved to the east and then stopped (I could gauge movement and direction by adjacent trees), then after 2 minutes it moved straight up. After moving up it began to appear more as a vertical line shape with three stronger points of light, one in the middle and one at top and bottom. It then moved down twice and then moved north over the next 15 minutes. In its last 5 minutes it appeared again like a blob of sparkling lights, became gradually dimmer and shifted to red and orange sparkling lights. It gradually disappeared over 20 seconds in a stationery position and did not appear again.

Witness #5

My wife, my son and I were watching TV when my mother in law came in and told us to come and look at these lights in the sky. When we got outside we observed the lights moving in an easterly direction. They were in triangle formation for the most part with a few others around them. They were a kind of reddish orange color and moved slowly across the sky. We could see a lot of them but in the direction they were going they would be out of sight soon, so I walked down the street. I continued to observe the objects until I was almost at the end of the street when one of them zoomed out of sight. And then they were gone.

It is the most amazing thing I've ever seen. I can't think of anything else that it could have been. Airplanes and helicopters do not move as silently and swiftly as these did.

Witness #6

I was returning from bird photography in S. AZ. I was driving between Globe, AZ and Superior, AZ at the time of the first sighting. I saw two amber lights in a location where I knew there were no buildings in the mountains. I was just about to enter the tunnel about 2 miles outside of Superior. I knew I would have a better view once through it. Unfortunately they were no longer there.

Only a few minutes had passed since the initial observation. I proceeded on and was able to now view a group of three lights that appeared to be between Florence, AZ and Queen Creek, AZ. I finally found a spot to pull off and view with binoculars. All that could be seen were amber/gold round lights. They did remind me a bit of the Air Force flare drops on the Barry Goldwater Range. The three lights went out, so I proceeded towards my home in Gold Canyon. Soon I saw two more lights and pulled over. I attempted to hand hold my camera with telephoto lens - the exposures were long and blurry.

I proceeded the last few miles and stopped in the Bashas shopping center parking lot where I pulled out a tripod and attempted photos/videos of three white non-moving lights. When observed with binoculars, they seemed to have a light orbiting the main light. I will review my results soon. All the while I saw the lights I believed they might be flares. I could never see strobes from a plane dropping them though. They did not behave like flares I have seen in the past. Plus from where I estimated them to be - they would not have been in military airspace.

I phoned a friend who lives in Queen Creek to go outside and look. He said they were to his west. I phoned a friend who is a pilot to see if he had a scanner or such to see if there was talk of this but he did not. Unfortunately, I had developed a bad enough migraine to not pursue trying to get a closer look.

I went home and told my wife. She helped me get back in the car and drove a mile or so down the road to a park area to try to see them. They were not there to begin with, but we finally spotted the last two of the evening very low on the horizon to the SW. She too, viewed them with binoculars. In every instance of sightings the lights were motionless but did seem to fade in and out. They were extremely bright on a very clear night.

I just felt confused and tried to rationalize that these were flares. If my estimation of their distance to me was correct, it doesn't make sense. If it was an extreme illusion as *to how close they were - flares would be my main guess. The white light objects I have no explanation for. Once again - reviewing the video and possibly having a videographer friend help me with it will be my next step.*

Witness #7
My wife, 2 sons and I were outside shortly after midnight setting off fireworks. My oldest boy saw an orange light in the sky and alarmed us. I asked him to run inside and grabbed our HD camera, but the battery was almost dead when I turned it on. We all saw the exact same thing, red/orange orbs rising into the sky from the direction of Phoenix Sky Harbor airport, one after the next, then changing to solid orange and shooting back and forth through the sky for 1-2 minutes until they faded out/reached outer space, then another rose into the air chasing a different direction.

This happened at least 3 times until it stopped, not sure how long it was happening before we went outside and noticed. They were very bright, stopped in mid-flight and changed direction, certainly not the airplanes we see normally, then continued on in another direction. My oldest boy says each orb was making a sound and he had dreamed this would happen tonight. I will upload what I caught on film to YouTube soon. We were all a little spooked.

Witness #8
New Year's Eve, my wife was out in the back yard with our kids lighting sparklers around midnight. She looked up in the sky and shouted for me to come outside. She pointed up at the sky and said, "What is THAT?" Both me and my sister-in-law got up and went out in the backyard and looked up at the southern sky, over our own house, and saw a red/orange glowing orb moving in a westerly direction toward the Gulf of Mexico. My wife pointed out the plane in the background with its blinking lights moving in an easterly direction and said, "It's NOT a plane, because THAT'S a plane right THERE!" There were NO sounds, NO flashing, just a bright reddish/orange glowing orb.

When it appeared to enter a cloud, it turned as black as the night sky, then when it appeared to come out of the cloud, it went back to its original red/orange color. Total time of the event was about a minute or two. My 8 year old got extremely frightened and started to cry after I answered my wife's question of "what was it?" with, "IT'S A UFO!!!" There were no zig-zags, no ascending or descending, just a straight flight path over the neighborhood that went out toward the Gulf of Mexico. Witnesses include me, my wife, my wife's sister, and our three children. No video or pictures were taken.

Witness #9
It was New Year's Eve and five of us were sitting around a bonfire. We noticed a round, orange object in the sky. At first we thought it might be an airplane but noticed two similar objects in a line directly behind the first. The three orbs then hovered for about two minutes and then one at a time each one dimmed and vanished. Then over the

course of about 15 minutes an additional 9 more orbs appeared the same way, traveling in a line, then hovering, dimming then vanishing. There was no sound whatsoever and there were 12 orbs total. We were astonished by the movement of these objects.

Witness #10

I am telling a story that was told to me by a family member. It is a truly spectacular story and I have a keen interest in the subject so I thought I would share it because I know they have not. Now I know these people and they are not ones to make up stories. They have no reason or anything to gain from this and they haven't told anyone besides family and friends.

So my aunt comes home from grocery shopping, it is dark out and they live next to farmland. She can see a good distance behind their house for at least a mile or more. She sees a glowing orange orb hovering around the trees a mile behind their house as she is bringing in groceries. She observes the object as she brings the items into the house, comes back out, and then observes the object as she comes back out to the trunk of the car. She gets to the back of the car, stares at the object wondering what it could be. She bends down to grabs some bags of groceries. In a short time - 2 to 3 seconds to grab the groceries - she looks back up and the orb had shot across the field in that time and seemed to be observing her. She panicked and ran inside to grab my uncle who had been napping. She brings my uncle out to observe what she was seeing.

They go outside and find nothing for a moment. They get to the edge of their property where the farm land starts when the object appears in the tree line where she had first observed it. They describe its movement as disappearing and reappearing in an instant. Erratic movements disappear and reappear from the bottom of a tree to the top of a tree in an instant. They observe this action for a minute when the object disappeared and reappeared and landed in the farmland behind a neighbor's house no more than 1000 yards away.

They describe the object to be as big as a house. It started to emit all the colors of the spectrum in sequence. It would emit an orange light, go dim and silently explode as they describe an orange light, then blue, red, green, yellow, all during the process I just described in rapid succession. At this point in a panic, my aunt started tugging on my uncle in fear to get them back into the house. At this point the lights stopped and they could no longer see the object. I thought this was an amazing story – I hope someone could share a similar experience.

Witness #11

My mother and I were outside taking out the trash when she told me to look up and asked, "What's that?" I looked up and observed a flame colored sphere, almost star-like. It was visible for a few seconds, disappeared very briefly and came back as two. A couple seconds later it vanished and appeared again, much farther to the east, as one sphere this time. Then it disappeared again and came back very close to its original position, this time it appeared to have some type plasma or material dripping or pouring from the bottom. At this point it appeared as two again, lasted for maybe 7-8 seconds and then vanished for good.

Immediately following the sighting I noticed what appeared to be a jet of some type flying directly to the same location. Over the next few minutes I observed other planes that appeared to be going that general direction as well. One thing I'd like to note, it never seemed to move other than phasing out and appearing in different locations. Seemed to blink itself miles across the sky in an instant.

Witness #12

It was about 2 to 3 o'clock in the morning when I couldn't sleep, so I thought I'd go outside get some fresh air. Then I started to notice a small pulsating light at the corner of my eye. I first noticed the light when it started to get bigger and brighter. When it started to cross the sky in crazy patterns, then more objects suddenly appeared. They began to pulsate sequentially while still getting brighter. After about an hour, all but one of the objects disappeared. Then the last object got much brighter and while rising swiftly to the sky started pulsating and the colors concentrated on the blues and greens (previously emitting a wide variety of colors).

While the object was flashing blue and green, I started to get violent headaches that lasted only about 1 - 2 minutes, then then the object turned solid white and shot across the sky. If I blinked I probably would have missed it going away. As soon as the object was out of sight my headache stopped. For a moment when I was having the headache I thought I heard a weird sounding voice. It seemed to be a language of some sort but the voice did not sound like a person. I could not figure out what the voice was saying. I THOUGHT I WAS GOING CRAZY. After all but one of the objects disappeared I was feeling very uncomfortable and sick to my stomach. That's all I can remember.

Witness #13

I was on my friend's porch at her home off of Rock Beach Rd in Irondequoit when we both noticed a large burning object slowly descending from the sky roughly 40 feet out and 100 feet up from where we were standing. We both immediately ran out into the street to better observe. My initial impression was that it was some kind of flaming debris but after only a few seconds of watching both the variation in the intensity of flame and the suddenly upward and outward flight path, I soon realized that what I was witnessing had no reasonable, earthly explanation whatsoever.

Believe me, I am a highly skeptical, maybe even pessimistic person and would not believe any of this if you told it to me, but it happened. We proceeded to visually follow the object by walking north to the end of her street. The street dead ends at the beach (Lake Ontario). We then walked east and began to notice at least 3, if not more of the same objects to the north and north east. By this time they were all at a much greater altitude than the initial orb was when we had first started to observe it roughly 10 minutes before. We did manage to get video of it for about 30 seconds when it was maybe 200-300 feet above us.

Once we were at the beach I noticed an immense structure slowly floating very high in the air. It seemed to have a kind of haze for lack of a better word that prevented us from getting a real detailed look at it though it was easy to track because it pulsed a faint light every 5 seconds or so. It's hard to explain but it definitely seemed to be a solid object unlike the red burning orbs that were more gaseous in appearance and moved and stopped erratically whereas the large object stayed on a somewhat straight path. The larger object completely stopped moving about 30 minutes into our observation and hovered at a very high altitude over Irondequoit Bay, barely visible if you hadn't been tracking it as we were. At certain points in time I believe I could make out a series of symmetrically spaced lights on this large craft and I also believe that I could see the smaller orbs seemingly buzzing around it but of this detail I can't be sure because of the high altitude and the haze that was either minor cloud cover or something emanating from the haze. I think the latter.

We observed all of this over the course of an hour or so and throughout this time there were quite a few visible airplanes that seemed coincidentally to have flight paths that took them close to these objects. My gut feeling is they were not commercial but military planes, but maybe that's just TV talking.

I would swear to seeing all of this unfold on a stack of bibles, under oath in a court of law and so would my friend. We immediately got online and saw some videos of red orbs over Lake Ontario that were very close to what we saw, except we saw the one orb from a close enough distance that we could actually make out flames emanating from or harnessed inside of its aura. An amazing yet unnerving night that I am going to try not to think too deeply about because of the scary implications of the existence of these things.

Witness #14

It was around 9 to 9:30pm when my wife said we had run out of diapers for my daughter so we went to our nearest TARGET store. It was a regular, clear sky, warm and humid night, typical of Florida. As we arrived, we parked our car and began to walk towards the store when suddenly both my wife and I stopped and stayed speechless. I as well as my wife were in such amazement and shock that we froze as we stared at the sky when she suddenly said "do you see that"? I gasped for air and said "yes". We didn't know what it was.

We saw a balloon like object, round in shape with a dark bright reddish color in the rim around it and as you looked towards the middle it was a glowing Orange/Yellow color that remained steady without change in color. My wife suggested it was a helicopter, I said no because 1) there was absolutely no sound and 2) it did not have a flash on the tail of what would be a helicopter. Then she said maybe it was a small aircraft, but again - there was absolutely no sound and it moved too smoothly and slowly to be an aircraft of our known experience.

All of the sudden a guy behind us screamed WTF! We are being invaded and took off running but my wife and I remained frozen looking at the object when all of the sudden another replica of that same sphere seemed to have come out of the one we were looking at - same size, shape and colors - then another and another and another till we saw a total of about 5-7. My wife said to me to film it with my cell phone. As other people stood in amazement, others panicked, so I did, I tried to catch a footage.

As the objects flew over us the distance from the ground up must have been that of maybe 100 floors up. The objects were seen very clearly and even seemed at times as if they were floating east bound very slowly. Suddenly the object as it continued its straight path moved towards the beach, (east) and the lights behind it that were in a formation of one on top another (lower on bottom) started to gather together connecting each other - kind of like if they were going inside each other until we only saw two lights and then they disappeared over the TARGET building. I heard people saying, "did you get that" but amazingly and weird my phone, like most others, had a black screen with no sound or image, kind of like if my cell phone was turned off.

We never understood what it was, we told our family and they thought we were crazy, other members said "you never know". Interestingly, an uncle suggested it could of been a paper balloon like the ones Asians make but if that was the case it would of flew up and not across, there was no wind, and it wouldn't have had the colors of glow we saw that day. My wife and I brushed it off for a while because we even thought we must be seeing things from our imagination, but about 8 months later as we were watching TV news briefs, we saw a news item that said "caught UFOs footage flying over the sea of South Florida. When we saw the images we saw exactly the same thing we saw that night and ever since then we became aware that we have seen something.

What did we see? We don't know, all we know it was very close, glowing, bright beautiful color mix, silent and kind of floating.

<u>Witness #15</u>

It was a Sunday night OCT 14, 2012 around 11:30 coming home from band practice from Hollywood going down the 605 south to the 405 south. I wasn't driving and usually don't talk much on the way home because it's a long tiring day. I looked in the sky to the right of me toward the ocean and what appeared at first to be a helicopter going slow toward the ocean. Suddenly it illuminated and pulsated a bright orange yellow and turned to the north.

The next thing I observed is the weirdest thing I have ever seen. The object was a white orb with four disc shaped objects connected to it with orange yellow lights moving around the white orb at mind boggling speed. Suddenly it turned all white, pulsated and when that happened a giant beam of white light came vertical to the ground and sideways. Suddenly time stopped and the object flew so fast a hole in the sky opened and I heard almost like a sonic boom and a red flash of light came from the object a half mile in each direction towards the ocean towards the car and all the light was sucked in to the hole into the sky and it was gone.

I felt like I had seen divine intervention. I was almost frantic. I told my singer that was in the car but he is a scientist and refused to look. But right after this sighting we somehow missed our off ramp to the 405 with a 20 minute time lapse. So we exited the freeway and I saw the white orb object with 4 orange disc circled around it. This time I asked my singer to look and he says "whatever" and this time numerous helicopters circling the object.

We got back on the 405 and I saw nothing until I returned home and here is what I saw. There are UFOs. I never believed in this stuff, but they are real. My sighting was an uplifting experience and I hope people in the world can see what I saw. I didn't have a camera on the freeway for the amazing thing I saw, but no one can deny this experience. THANKS EVERYONE - LIFE IS GOOD!

Witness #16

I was driving home with a friend. He was the passenger, I was the driver when I first noticed very bright orange/red lights glowing above the tree line. What made me take notice at first was their color, size, and brightness. At first I didn't know what to make of these lights so I pulled over.

Both my friend and I were in shock and awe as to what we were seeing. Quickly we got back into my car and drove not even a tenth of a mile to get a better look. It was then we saw how many of these lights there were. Almost immediately after I pulled over, another vehicle pulled in behind me. A woman got out of her car. I remember her saying, "what are those?? Those are not airplanes." Seconds after the first lady pulled up, a man had pulled up to see the lights as well. He had a small child in the back which I noticed before we all left. My friend, the lady, and the man that pulled up behind us all just stood/sat there in awe as to what our eyes were actually witnessing. Big, bright, orbs, orange-ish/red in color, 25-40 of them. Extremely bright, moving all in the same direction, at a very slow speed.

I know what airplanes or helicopters look like, these were not them. The lights did not blink nor flash. They were undeniable. More than obvious that they were there. It almost looked like a convoy. They made absolutely no noise at all. Slowly there weren't as many orbs, so I asked my friend who was with me, "where are they all going?" when I realized they weren't going anywhere.

They dimmed out, as if they were on a dimmer light switch. They just faded into the night. After about 3-4 minutes all lights had faded out. When I first saw the lights, I have to admit, I was extremely curious. Then once I had the better view of how many and were able to observe them, I was scared.

I have never seen anything like this ever in my life before. I was in shock, I was excited, I was scared, I was dumbfounded. I was able to snap a quick photo using my camera. However, it does no justice as to what we all saw that night. It is hard to see the *orbs, and I was only able to capture seven. They were not all in a line like that. Some lights were above others, some side by side; however, they were all traveling together. Also the brightness is nowhere near captured in the photo either. However, we all know what we saw that night and what we saw is the unknown.*

Witness #17

Witness stated: My partner and I were swimming in our pool at night when all of a sudden a giant glowing orange ball appeared 100 feet above the ground. It continued at a decline of approximately 45 degree angle. Had spike like shards. No sound. No way was it a balloon.

Witness #18

Some soldiers and I here at Fort Sill were getting ready to go on a brigade run. I felt a strong urge to look at the sky for some reason, a feeling I usually do get before a UFO is nearby. Soon in the direction I was looking, I saw a dull green orb that seemed to come from out of nowhere off in the distance. You had to have been looking in the exact direction it appeared or you'd have missed it. It descended slowly, dulled out almost entirely, then it swelled up and glowed the most brilliant green you could ever imagine, stopped a moment as it did, then it began to speed up and shoot downward with a glowing golden-yellow flame with a reddish orange outline with a tail, almost looking like a meteorite burning up. But the way it descended and then suddenly stopped was very peculiar.

It was visible by many soldiers that were standing in a brigade formation and many of us agreed it indeed could be an extraterrestrial presence. It only lasted for 30 seconds and no sound was heard when it supposedly crash-landed and we thought nothing further of it. It most definitely landed on Fort Sill though in the area where artillery fire or field training usually takes place. However there were no field training exercises or test fires going on at the time and we have no weapons that behave like that. For those of you that wish to email me on the matter feel free to at [Email removed/cms/tg]. The exact number of witnesses is unknown but I can confirm three other soldiers I know personally.

Witness #19

I was sitting on my back patio watching and filming wildlife and observing the clouds. I saw a glowing sphere coming from the East moving toward the Northwest and upwards very slowly. I grabbed my camera and began filming. I couldn't see the object in the view-finder so I only got a few seconds of footage. The orb appears much smaller on film than it did to my eyes. After one minute I grabbed my binoculars to get a closer look at the object. I have drawn what I saw. I have never seen an orb in the daytime and have never seen anything like what I saw before.

I am feeling extremely disturbed and perplexed. I don't know if what I saw was a machine or organic or some sort of unknown entity. I watched it for 5 or 6 minutes until it disappeared into or out of the upper atmosphere. Sequence: 1) main orb appeared. I tried to film it. 2) Decided to look at it with binoculars. 3) Observed sphere pulsating colors. 4) Observed a small group of white specks above sphere. 5) Observed 3-4 small orbs dropping from large sphere. 6) Observed white specks above sphere elongate upwards. 7) Lost sight.

Witness #20
It moved in a straight line. No visible lights on outside, bright orange around outside darker in the center like a doughnut.

Witness #21
I witnessed this sighting around 2am on a Thursday night in Myrtle Beach, SC while I was staying in the Wyndham resort/hotel on vacation. I was up in my room outside on the balcony at the 12th floor which was pretty high up, so I looked out towards the street enjoying the view and noticed in the sky - not up in the sky it was more straight ahead from the height I was already at maybe a bit higher than I was - a bright orange ball pulsating like a dimmer switch on and off. The direction it came from was the north and I observed it as it moved silently across the sky towards the south running parallel with the coastline but on the land side not the ocean side. I eventually lost sight of the object as it left my field of view.

At first glance I couldn't figure out what it actually was but then came to the conclusion after gathering myself from being in shock that it couldn't have been anything else except a UFO! Not a plane, helicopter, balloon, satellite or star...I know what I saw that night.

Witness #22
This is our 4th positive sighting of these orbs in the past 3 years with 3 of the 4 seen over water. At 7:45 pm we noticed the orb hovering around 500-1000 foot above Safety Harbor and emitting a red cone shaped beam of light down at the water. We live very close to several airports both private fields, international, Coast Guard and an Air force base so we are very familiar with air craft and helicopters of all shapes and sizes. This orb had no navigation lights - only an amber glow masking its actual shape and size.

I have found these and other craft are attracted to green lasers and actually carry one with me and once I confirmed the object was not an aircraft with required legal lighting I pointed the object with my laser. The object appeared not to like the attention and turned red for a second and then went dark but the object was still there for a couple more seconds as my laser would reflect off its surface. Then it disappeared in an unknown direction.

The orbs are always silent and seem to be trying to camouflage themselves as conventional aircraft, stars or lights. I personally believe they hide in the water during the day and come out from dusk till dawn.

Witness #23

On my way to work at 4:30 am. Standing in front of Disneyland I noticed under the cloud cover a bright red pulsating object. Moving slowly west and then making a horseshoe turn. The object than proceeded to "shoot" straight down a smaller red object. It then, still in full view under the cloud cover, went out or just vanished. I was fully awake and aware when this occurred. It was clearly not a conventional aircraft. It was fluid or plasma like, the object. I am on the bus filing this report on my way to work only an hour since the sighting. My feeling right now is being irritated. Sooner or later someone will have answer for these sightings.

Witness #24

I was driving home @ around 10:15 p.m. last night 1/1/2012 on I-75 heading northbound. When I was 1 mile south of the Miami Gardens Drive exit I looked up and there were 10-13 red orange orbs hovering stationary over the road about what seemed to be 100 miles up. At first I thought they were red paper lanterns, but it was clear to me after observing them for several seconds that they were not anything common. Others had pulled over to look @ these things. I wanted to call the FHP office in "Plantation" to report them and to find out if anyone else had seen them. Instead I came online to find where I could report this sighting. Please let me know if anyone else has reported this. It happened and it was real.

I am a healthy 35 year old male with no history of insanity or psychological issues. This was real. This is the first time I have seen this with my own eyes. What is going on here? If there are government experiments or actual UFOs we need to be told. This was real. I do not have a cell phone so I could not take any pictures, but others had pulled over to look at them, so I'm hoping they had taken pictures of these red orange orbs. I hope you have received other reports such as mine. Again...this happened last night @ around 10:15 p.m. over I-75 (northbound) 1 mile before "Miami Gardens Drive" exit. I wrote the number of witnesses to be 7. There were others pulled on the side of the road but I cannot give you an exact number, I did not pull over. I drove in the slow lane looking up at these orbs and noticed other cars pulled off to the side and people looking observing.

Witness #25

On 9/8/2011 at almost 8:00 p.m. I was working on an addition to my home. I was on my front porch and out of the corner of my eye I witnessed a flash of bright reddish orange light. I turned immediately and saw what I could only comprehend as a fireball of a size that was unbelievable and the light that came from it was of an intensity that was frightening. I witnessed an object that appeared to come out of thin air, it left a short trail of light as it appeared, then it stopped and glowed, and pulsated with a light that was so intense and powerful that I was expecting an explosion or something worse. The event lasted approximately 10 seconds or less but what has kept bothering me since the event was when it disappeared the object left a trail just as it did when it appeared and it was gone. It disappeared into thin air just as it appeared out of thin air.

I have replayed what I saw in my mind and it has plagued me since. When the object appeared, stopped for what was seconds and pulsated with a light that was of an intensity that was not natural, it was a light that was of immense power and literally scared me to a point that I expected an explosion. The object was controlled. It has taken me some time to comprehend what I witnessed, at first I was so frightened that I contacted the Charlotte County Police and several officers responded to my home. I told them what I witnessed, at the time I could only describe it as a fireball meteor. I also called channel 13 news to report what I witnessed in hopes that others had seen it too. I was worried that something bad had happened, I now no longer believe what I witnessed was a meteor, it couldn't have been, it just didn't have the appearance of one as it stopped and pulsated.

I also reported it as a meteor sighting to the American Meteor Society at the time and in the back of my mind I felt what I saw was not a meteor, but I could not comprehend it to be anything else. Since the event I have felt a sense of impending doom, what I witnessed was not natural. I feel I witnessed a power greater than any known on earth and it wasn't friendly, I feel it was a probe of some sort, a presence of immense power and destructive capabilities, I am sure that what I witnessed had intelligence and intent.

Witness #26
My brother and I have been seeing these UFOs every night. The first thing that makes them so very easily noticeable is it looks like a star that is flashing every color in the spectrum. I thought right away it was a UFO because stars don't flash and pulsate many different colors. It is a pulsating object that looks stationary with the naked eye, but when in the camera lens, it is moving at light speed - it moves so fast it looks like it is at a standstill.

After taking several photos, you can see that they are doing speeds and maneuvers we don't have the ability to do. You just have to see the pics and come to Gerald and look in the sky and you will see what I am talking about. I want to know what they are, it isn't something that we should take lightly. Please contact me and when you see the photos I took I'm sure you will definitely want to come see for yourselves. The pictures are 100 percent legit and evidence there are many UFOs everywhere and more are present every day!

They scare me and make me very uncomfortable because you can tell they are here to stay, for what I don't know. After me and my family watched for a while, I saw another one come and appear close to the one we were watching and well - after seeing more than one, my niece got a bit scared and went inside. So we lost sight when we went inside the house.

Witness #27

We saw this glowing star like ball hovering over the (Pacific) ocean. Then it started to go down like the sun, it was so bright it started to hurt my eyes just to look at it!! It also left a glow on the ocean water like the moon does on the ocean. It started to change color like reddish and orange and a glowing bluish white color. The light got bright and then it seemed to dim. There were a lot of aircraft in the area and boats in the harbor. I would think someone else saw this!! As it ascended down and appeared to be moving farther and farther away, it turned redder!! Then it was like a little red dot than disappeared!! This is the 3rd time I've seen this object this year!!

Witness #28

At around 7:20 pm on November 17th, southeast of Deputy, Indiana I stepped outside of my house and started walking towards my vehicle. I looked behind me in the sky and there was this big bright orange/yellowish ball that would slowly get bright and then slowly go dim. Sometimes 4-5 smaller balls would shoot off beside it and remain in a straight horizontal line next to the bigger object (not falling to earth). Then all of the little ones would return to the bigger ball. Each time the light became brightest or basically became visible again it was followed by a far off rumble similar to hearing a far off thunderstorm.

This all lasted about 10 min and then I never observed it again. When I first saw it I thought it was a shooting star or asteroid but it and all the other smaller lights were stationary. When observing it, I wracked my brain thinking what it could be. I thought maybe a helicopter doing search and rescue with a spotlight that I could only see when it faced my direction but that doesn't explain the far off explosion or rumble that I heard at least 3-4 times. The object just never lit up again after about 15 min.

Witness #29

I live in Scranton, Pa. in an apartment house on the 3rd floor with 3 big windows where I can see a lot of the sky and the area the windows are facing towards Wilkes-Barre Pa. And yes, I know there is a airport there and I lived right next to it, so I know the difference between planes landing and taking off.

This morning December 3rd at 6:30 I woke up to something in my head saying to look up in the sky and I did just at the wrong direction but then noticed something in the sky flickering in different colors. At first it was clear and seemed close so didn't think much of it and then I turned back and saw it move to the side a little so I used my window frame to see if the object was moving - and yes it was, at first really slow to the right so I knew in the matter of 15 minutes a star cannot move by that much and it wasn't a helicopter or airplane – I am sure of that. It then started moving a little faster. I even woke up my mom to witness it and she did see it but not as long as I did because it went behind a big tree that is there and that's where I lost sight of it, but I have never seen something like this where it gave off different colors and made me feel all happy to see it.

I have seen stuff like this before, but never with flickering colors like this. It went blue, green, yellow, and red - and not like underneath a plane as a whole sphere giving out the colors. I just wish I had my camera working, but could not go without at least reporting it to see if someone else maybe saw it. I am sorry for the spelling and bad English. I am originally from Eastern Europe but I swear this is no joke and I just want to find out more!

Witness #30

It was the 4th of July and I, my wife, and my granddaughter were in the yard lighting small fireworks for my granddaughter's amusement. My wife suddenly said to me "what in the world is that"? I looked in the area that I was referring to and a reddish/orange colored sphere was coming in and out of the clouds. At first I thought it was a weather balloon until it started to change shapes. It changed to a tube and started to tumble then regained its shape as a sphere. Thinking quickly, my wife grabbed the video camera and started to film the event. She was able to obtain about 10 minutes of footage before the object disappeared into the atmosphere. My wife's son claimed to have seen the same object at his apartment and took the camera with him and he was able to collect quite a bit more footage of the same object that we had seen at the house. We still have the footage and would like to have someone take a look and see if they can explain what this object could be. We do not have the technology to upload the video, or else we would attach it to this statement. The video is on an 8mm cassette tape format. (3 witnesses)

Witness #31

I, my wife, and two of my friends witnessed this event. We were all in/around my friend's backyard pool. My friend Anton first noticed the object heading towards us from the south. It was traveling northbound along I-35. Anton spotted the object on the Southern horizon, so we got a good look at it as it flew all the way from the Southern to the Northern horizon in about a minute. The house is right off of I-35, so it flew right over us.

At first it just looked like a black sphere, as it got closer we saw a constant orange light on the bottom (not blinking). The object was flying at approximately 2000 ft. It was going about the same speed as a passenger jet, but was completely silent. We are in the landing path of the Des Moines airport and there are frequently jets flying at this altitude, and it is always very loud. The object appeared to be about the size of a two story house. The object moved from the Southern Horizon to the Northern horizon and out of sight in about one minute.

Afterwards, we all tried to rationalize what it could have been. My friend's dad was in the Air Force for over 20 years, so we called him and told him what we saw. He said it sounded like either a secret government aircraft or a UFO.

Witness #32

I was taking a motorcycle trip along Route 66 and I was riding my motorcycle to a fellow professor's cabin on Lake of the Ozarks Missouri to stay the night off the Route. I got the directions wrong to his cabin and was heading West on Highway 54 at about 11:00 p.m. CST right on the outskirts of Camdenton Missouri when I noticed a large sphere on the left hand side of the highway above a commercial building (it was South). I thought it was a large dealership type balloon at first, but it started moving very slowly across the highway in a zig-zag style fashion. I immediately slowed my motorcycle speed down to about 10 mph, as I was transfixed on the object and could not keep my eyes off of it.

I thought it was maybe a hot air balloon ride as I got closer and thought it was amazing that they had hot air balloon rides so late and so low to the ground in the Ozark hills. It was a spherical type object but was not static in shape...it was hard to identify what it was for some reason. I couldn't see any type of hot air balloon basket, but for some reason I felt as if I was being watched while riding on the bike. I kept wanting to pull my bike over to take a picture, but was too amazed looking at the object that I didn't for some reason.

It was large (the size of a small house) and floating about 150 foot above the ground...it was reddish colored with a glow (I thought the glow was the fire for the hot air balloon). It passed right over the top of me along the highway and was hovering along the right side of the highway (the North side). I was trying to find a place to pull my motorcycle over so I could take a picture, but trees were getting in the way - but I could still see the object for another 10 to 15 seconds or so. I finally found an open spot with no trees to take the picture and within seconds of getting off my motorcycle, the object faded extremely fast heading in a 45 degree angle. It was literally gone within a few seconds, but there were no obstructions or trees for it to disappear behind. It couldn't have been a hot air balloon because of the exit speed and how low to the ground it was. It was definitely sphere-like.

The whole experience lasted only a minute or two, but I have tried to rationalize what I saw almost every minute of the day since then (It was 6 days ago). The professor's cabin who I was staying at knew what I had seen, as he had seen the exact same type of object hovering above Lake of the Ozarks 3 years ago in late fall 2006.

We are co-workers at a University...both professors...I am a Law professor and he is a Marketing professor. He told me that I could come forward with his information as well. He and his wife both witnessed the object hovering above Lake of the Ozarks for about 5 minutes...they could not believe their eyes...and were also so transfixed that they didn't capture it on film. He informed me that he kicks himself all the time for not taking a picture or video, but for some reason he was so mesmerized he didn't move. The fellow professor and his wife never told anyone but relatives in the past...but he and I have talked for hours in the past 6 days to try to rationalize what we saw, but there is no rational known explanation of the object, how it moved, its size, shape, and aura around it.

I have never been an avid UFO follower and was not looking for a UFO...it found me...and I completely believe and now intend to spend my summers investigating this phenomenon (as a professor my summers are fairly free). I will help you in any way I can...I am 100% CERTAIN that I witnessed a UFO...as a lawyer and law professor I am very skeptical without evidence...but the evidence was literally about 150 feet right above in front of me...it was the most amazing thing I have ever witnessed in my life.

I will remain anonymous for now, but if for proof or other purposes you may request me to come forward, I may at some later time. I also plan to get a tattoo of what I saw on my body...I do not have a photo, sketch or audio or video as of right now, but I intend to have one created for my tattoo.

Witness #33

This is something I have sighted numerous times over the last 18 mos. The first time I sighted it was the clearest, but every time after I could tell that it was the same anomaly. The first sighting was at approximately 12 midnight in the month of July 2007 in the Burntwater area of the Navajo reservation of Arizona. I typically watch the skies in the evening due to other things I have seen which seemed strange to me and it has become of great interest to me.

On this night I observed an object that appeared star-like but much brighter and way too close to be a star. After watching for a few minutes I noticed that it got brighter and then dimmed a bit, but the strange thing was that there were multicolored lights flashing in a definite pattern around the perimeter of the object.

To say the least my curiosity was piqued, so I brought out a telescope to get a better look. This was the amazing part - upon observation with the telescope I saw a perfect circle with lights - some brilliant green, some appeared amber, bright red, white, yellow, and vivid blue. The really strange part of the sighting was that the circular space between the lights appeared to be nothing but a void of blackness. The object stayed in this position for a couple of hours and later that night it appeared in a different location and once again stayed there for some time.

This was not my only sighting of this object and one night at the same locale there were three of them in different parts of the sky. All were highly visible and the pattern of lights were visible with the naked eye.

I also spotted this anomaly over the Gila wilderness of New Mexico as I am an avid backpacker. Recently in Oklahoma where I now live, the object has appeared numerous times and the object moves in an apparent random pattern and not as in an orbit or as the movement of any celestial objects I know of. I have shown others this 'thing 'and they *were just as surprised as myself. If any other people have observed this phenomena I would love to hear about it and compare notes.*

Witness #34

I was leaving my girlfriend's house around 10pm. I usually take my boat out because I live on a lake and our houses are close by. It is about a 10 min boat ride. As I was leaving, I started the boat and was heading east to my house. Something caught my eye to the southeast of my heading. My first instinct was that it was a planet in view maybe Venus or Mars because it was glowing. But I took astronomy in high school and it was too low and too bright.

I am a man of reason, and to me everything has to have an explanation. I stopped the boat and looked closer, then when I did stop it was changing colors to green, red, and white. At this point I thought it was a plane, because there is a local airport nearby about 1 mile from my location over the trees. At the time the object was about 3 football fields away and 4,000 feet up. I started the boat again and was moving east at about 10 mph. I kept watching it. At the time I called my girlfriend telling her what I saw, she thought I was crazy, I told her to come out and look at it, but she hated the dark and said it was too cold, so I hung up.

I am an emergency medical technician at a local hospital and people take me very seriously, so when I tell somebody what happened this night they believe me. I stopped the boat again, and the object got closer and dropped altitude to about 1,000 to 2,000 feet, and 200 yards or so. I was about 5 minutes from my house. I was getting scared by this time. As I stopped again, the object was now closer about 500 feet up and 100 yards away. It seemed when I stopped and started, it got closer and lower, I knew now this was not an aircraft or star or planet, it was definitely a UFO.

This time I could see it very clear, no sound, medium sized object sphere star like shape, changing colors every 4 seconds, to red white and green. After the 3rd time I had enough, and floored my boat back to my house and ran inside. Before I went inside I looked back and the object was gone.

I never believed in UFOs or aliens but after this September night, you can bet I do now. I never do drugs drink or nothing to conflict with my wellbeing and this is the first time I told nobody about it. I used to go on the lake at night, now I don't go out at all unless imp with somebody. I am 21 years old, and I have never been so scared in my life.

Witness #35

This was a few years ago so some details are a little fuzzy. It was Halloween and I was trick or treating with my friends in Williamsburg. We noticed a glowing orange light in the sky and stopped to watch. It was soon joined by 2 or 3 others. They then proceeded to drop this glowing orange substance, which trickled down and looked sort of like a flare. We watched them do this for a few minutes, and I'm not sure what happened afterwards. We either just walked away after a while, or they disappeared and then we walked away. I thought they might have been those lanterns that people light off on the 4th of July, but I don't know why someone would have been doing that on Halloween. Also the objects behavior indicated otherwise, as they appeared out of nowhere and hovered in place.

Witness #36

I am kind of guessing the date (I can find out the exact date however because we were coming home from a church function). I was heading East on Chase. I looked to my right (south) and up above what you would call "Mt Helix" was a glowing red light. I opened our HUGE sunroof in our Forester (3x3ft!) and we were stopped at a light for a while ... I watched. It seemed fixed in location. My husband thought at first it was moving until we stopped and he looked too and saw it wasn't moving. It faded for a minute then came back. If anything, it might have been gaining altitude as it seemed to get smaller and dimmer ... but then came back. I swear it looked like someone had lit a candle inside a red paper bag ... or orange ... kind of cross between the two. It wasn't huge but it wasn't small. It wasn't that high up either. I kept my eye on it telling husband to HURRY home so we could get binoculars.

We got home (about 1 min away ... about 1/8th mile away) and ran back out. By this time it was quite dim but again came back. Not really changing positions but again seemed to come back down lower a bit or at least get brighter. It was never BRIGHT like a LIGHT really ... it seems more "organic" if that makes sense ... It didn't blink or anything but it "flashed" or "flickered" I guess is a better way to put it. It would both get brighter and dimmer. There was at one point some slight movement. Almost like it was being jarred. If I could guess I would say (from what perspective I did have which wasn't much) that it may have moved 15 ft. in any direction pretty quickly, again like it was being bounced around or jarred. Almost like it "wobbled". It would get quite dim when this happened. It was just slightly foggy or maybe just a bit cloudy. As it got dimmer, and more "flickery" I had a hard time telling if it was "going out" or "leaving" or being covered by fog/cloud layer.

I know for a fact that other people saw it because I saw other people looking out their windows on the drive home. It was VERY obvious in the sky. I was really surprised that I didn't hear anything about it. I once saw a blimp (my first live viewing of one actually) come from very far away that was lit up on the inside ... it was a Saturn blimp so it too was red ... and I thought it was a UFO at first until it got closer of course. My first thought was that was what this was and it was just moving away from us. But because of its movements and behavior that was totally ruled out, especially after having watched it for so long.

In total from the time we first saw it, until it was out of sight, was about 15 minutes! I watched the news but didn't hear anything. I was too afraid to ask anyone if they happened to see anything.

Witness #37

We were at Oak Island on the beach fishing when my wife actually noticed them first, when she saw them she immediately pointed, asked me what it was and I turned to see, in the southwestern sky, at about 2 o'clock in position, there were two bright red lights. One was very large, oval in nature, dare I even say saucer-like in shape. The other was tiny and round. The larger one would make maybe 20 to 30 of the smaller ones in size and was stationary. But the smaller one was to the left of the larger, moving down and to the left, it then circled back up and to the right, back towards the large object. Once it reached the big one, they appeared to merge and then the large one just dimmed out really fast as if someone turned off a switch. As it was over the ocean horizon I couldn't swagger an estimation of how far away they were. From what I have found out so far they were a little to the left of a star called Arcturus and a little above it in elevation.

There was no sound and the entire event maybe lasted 15-20 seconds from the time I turned around. This was at 9:20pm on the night of Sept. 19. We didn't have a camera with us as it was night and we didn't think we would need one. We were both awestruck as we have never seen anything like this before and have no idea what they could have been. If it was a man-made aircraft I have no idea what kind it could have been as these did not act as a typical aircraft would.

Witness #38

On July 05, 2013 I was standing in my driveway when I saw a flash in the southern sky. I looked up to see several red orange spheres going from south to north, then abruptly change direction and head due east there were approximately 30 to 50 of these that I saw over about a 5 to 7 min period. I yelled at my neighbor and tried to get my girlfriend to come quickly to observe the objects. She just thought I was crazy so I am not sure of how many others may have seen what I did, but I can tell you that I watch the sky a lot closer now then I use to, thank you. PS I took pictures with my cell phone and none of the pictures showed anything but black sky.

Witness #39

I was getting ready for bed when I glanced out the window and saw a light in what I thought was a strange place in the sky for a star or planet - that was around 9:30 PM. I watched it for a few moments and saw that it had shifted positions from where I first marked it. I then got out my binoculars and went into the back yard for a better look. Sadly, they are only 10 X 25 and did not offer as grand a view as I would have liked. I brought it to the attention of my girlfriend and after her assessment she took out her camera, which did not offer much better resolution, and after only a few attempts at photographing it, she gave up.

Through the binoculars and the bedroom window I continued to watch the lights change colors and shape. There were green and white lights at what would be the north end of it, two green side by side with the white one on top, while the middle strobed red, yellow, pink, blue and white. As it floated across the sky ever so slowly, the lights at the north end changed positions and began to glow bright white at the south end without the green ones, while the lights in the middle seamed to change from a concave shape on the bottom, to seemingly encompassing it around unseen edges.

The lights contained within the middle of the object were my main fascination and changed many more times. In the beginning they were round and only outlined the bottom, then encompassed it wholly, in nearly a perfect circle, flashing a multitude of varying colors, leaving a dark, or, black stripe down its center, (north to south.) At last glance there were no lights at the north end and it seemed to have a "C" look to it as if it were heading south with the bright white light heading it up. As this is still happening at this moment and I have taken yet another look, the bright white lite is now on the north end, while the surrounding lights have changed from nearly horizontal, to nearly vertical. It is now 10:30 PM and the lights are still moving in a southwestward manner at about the same pace as the other stars, which are NOT, flickering multi colored lights.

If this is a planet whose light is flickering in the clear and cool night sky because of the atmosphere, then I have to say, I can't wait till tomorrow to see it again!

Witness #40

Smoking a cigarette looking into the starless dark clouds when I noticed a faint globe above my neighbor's house. It was moving at incredible speed and was almost see-thru, changing all sorts of colors. Making drastic movements as if it was unaffected by wind or gravity. It was only visible for 20 or 30 sec, but I can't explain what I saw.

Witness #41

March 26, 2012: My friend and I came out of my house to take my dog for a walk. It was just after 9:30pm, we were walking later than usual. We stopped at the sidewalk and were looking at the moon with Jupiter and Saturn all in alignment. We were looking west. Over the two story house in front of us came a glowing orange ball. It was about 70 feet up maybe. It moved very slowly and was as bright as a yellow/orange streetlamp. I said: "What the hell is that?" as we looked up at it.

My first thought that it was a helicopter and that there was a fire in the cockpit. I thought it was going to crash. There was no sound and it did not emanate any light. It just glowed like a ball of fire, the surface was like moving molten lava, the edges fuzzy and flaring. It was about the size of a basketball. It hovered right above us as we were facing south. It stayed there for a couple of minutes, then continued to move east. The dog didn't react at all to it and we were not afraid, just puzzled.

At this moment I thought I should document it, so I dashed into the house to grab my camera (unlocked door, opened camera case, ran out, turned on camera, focused, and snap). By this time it was considerably smaller and was heading south a few streets over. It looked much redder than when it was close. After I took the photo, it warp sped away.

The next day we were at a funeral. Afterwards we were telling a few people about what we had seen. Another neighbor who lives on the west side with an ocean view said that she had seen an orange orb hovering over the ocean at about 9:30pm, wondered what it was, but she did not get a close up view as we had. My photo was taken at 9:42pm (on photo it says 8:42 but had not allowed for daylight savings).

During the sighting I was puzzled, but calm. After the sighting I was a bit scared. We continued to walk around the block. After returning to the house another friend knocked on my door. She said my eyes were as big as saucers when I opened the door. I have always enjoyed looking at the sky, I love physics, and can explain most things. After this sighting I was jumpy and a bit freaked out by it because it was something unexplainable. My friend reacted more nonchalantly and wants to see it again. Easter Sun 9pm another neighbor saw orange orb over same street.

Witness #42

A friend and I were playing a video game when we stopped to go outside to have a cigarette. As soon as I stepped out the door I looked up and pointed to my friend and said "what the hell is that"? My friend said that it's probably a UFO and laughed, the object was bright white towards the center with red, blue light pulsating from the center. The object remained visible for about 2 minutes, then just vanished like someone turning out a light switch.

It was not an airplane as we could see planes flying high above in the skies. The object was around 35 degrees in the air towards the southwest of Republic, MO. I would estimate that the object from my viewing point was the size of a half dollar. Hwy 60 west and east run right where this object was and traffic could be heard from the location of my home so someone else had to of seen this. It was out of place in the night sky at 8:33pm.

Witness #43

I saw a light in the sky that seemed like a star over the hills in Pinole, CA. It was 11:00 at night and I was outside smoking. The star seemed to have a colored strobe on it but it wasn't moving like most aircraft and remained mostly in place except for motions laterally not vertically. I told my wife as I went into the house and got my telescope from the back bedroom. Once I got the object in the viewfinder it became apparent it wasn't a star or a plane. It was clearly a sphere and it radiated a light of all colors; yellow closest to the sphere and red further out, ending in violet. The light were not just light but seemed to be more colored spikes than a light hue from other stars or the moon. It moved back and forth in a short pattern like a right triangle. My wife looked through the telescope and confirmed what I saw.

We watched it until it left our field of view around 12 a.m. We talked about it and both decided that what we saw wasn't a plane or helicopter or a star. We agreed on the shape and the colorful spikes, and we agreed it was nothing like we had ever seen before. The next night at around the same time it was in the sky again except it was exactly 180 degrees from the first night, meaning it would have been exactly behind me from the first night. It demonstrated the same behavior until disappearing.

CHAPTER TWO
WHAT DO ORANGE ORBS DO?

Witness #1

My husband and I were sitting on the balcony of our condo on Orange Beach in Alabama. It was approximately 9 pm. We noticed in the southwest sky a large orange orb-like light appear over the water. It was very large and very high. The light multiplied into 2 or 3 more lights - I'm not sure because at this time another light appeared to the left of this one and multiplied into 5 or 6 evenly spaced lights in a straight line and unmoving. It was so quick and shocking to see. Then they disappeared. They could have been 10 miles out over the gulf- but they were mixed in with the clouds which were spotty. They made no sound at all. They seemed like flares or Chinese lanterns in appearance but did not act like either one. The one that lit up in a straight line seemed to all be connected. We have observed Chinese lanterns over our house before and these were different. These would have had to be released from a boat far out in the Gulf.

Witness #2

There were 10 of us on my back deck. We were watching the fireworks. We saw an orange round object zoom across the sky away from the fireworks. Then there was another one and then two close together, then one. They continued to fly across the sky away from the fireworks in a straight line. They got to a certain spot in the sky and disappeared. We are not sure if there was a cloud they went behind as it was dark.

When we first saw it, we thought it was part of the fireworks display, but as we watched we saw that this was far from the case. There were approximately 15 of these objects following each other. They were not closely following. It would be a minute or so before the next one came. The whole procession took about 15 minutes. The fireworks were over as they orbs continued their path. In fact they went right through two fireworks that were in the sky. They continued their path as if not affected by the lights. My sister-in-law tried to take a picture but all she got was a blur. When she tried to enlarge it she only got a huge orange blur.

Witness #3

I was waiting for my buddy to come over. He called me and told me to look out the window right away. I went outside and saw about 15 or more lights dancing in the sky. Couldn't really tell which direction they were headed. They were changing formation and then just kind of disappeared up into the atmosphere. It looks like they communicate with each other by blinking and movement. As time went on some were disappearing, getting smaller and changing to a more maroon color. No clouds were in the sky at the time, but minutes later they left clouds behind like weird spotted twirling shapes.

Witness #4

Okay, the adrenaline is still flowing, but I want to write this down as it's fresh, forgive my grammar. Driving north on 680 in San Ramon California, saw 5-7 red-orange orbs in the sky flying in impossible formations. They were able to rapidly reorganize into different flying formations. I turned the car around and exited the highway. All four people in the car witnessed this. When I found a safe place to try and film I could not get around the trees and buildings to get a clear shot. Another driver pulled their car over and witnessed the same thing, I gave this driver my binoculars to try and see if he could see while I tried to capture on my iPhone.

These things were fast and I just couldn't catch them on my iPhone, I really tried. I cannot express how intense a feeling it is to see a real UFO and trying to get a good look. The objects basically started to fade out one by one and then they were gone. I saw these same type of objects on May 6 in the Bay Area. I have no doubt something major is going on. We were seeing their glow, there has to be a solid surface, but the distance was too great to discern. They were at around 20K feet by my estimate. There is no way at these were planes or helicopters as they move way too fast and change direction in impossible ways.

Witness #5

Observed orbs form into one and attach themselves to mother ship. There were several orbs in the sky for long periods of time, but they moved quickly through the sky. The mother ship hovered for long periods of time. This is the second sighting in the last few days in Woodlands Texas which I have observed in the late evening hours. More have also been reported by other individuals in the area.

Witness #6

I saw 10 or more orange/red orbs/balls of light flying in formation changing directions and forming patterns like triangles. I also saw what looked to me like the Big Dipper in one of the formations. There were about ten other people with me at the time all of us saw the same thing. It lasted about 20-30 minutes.

The objects cited above remained stationary for approximately 10 minutes in a constellation type pattern. All the while they were emitting an amber/orange light. We pulled over to observe, and 6 or 7 objects moved and quickly disappeared. The remaining object remained stationary for about 3 additional minutes and continued to emit orange/amber light. It then moved abruptly to my right (south?) and suddenly disappeared.

Witness #7

The time was 1:20 am when I happened to notice from my backyard at approximately 35° to my horizon, a dim orange orb looking northeast of my position. It was faint at first, but grew brighter and appeared to be going faster headed northwest. The stranger part was that quite a bit back behind it I saw a flashing strobe-like light appear out of nowhere and looked like it was chasing it. It gradually started catching up to it as they both sped up! Both the steady glowing orb and the flashing light traveled from northeast in a counterclockwise circular pattern all the way around across the sky's horizon in a 3 quarter circle radius. After about 5-6 minutes of watching these lights travel across the sky in a circular pattern the flashing light that was behind the other got up very, very close to the orange orb light and suddenly stopped flashing and was gone. The other orb light kept traveling, but I lost it as soon as it got real close to the moon due to the full moon's brightness. I have never seen anything like that and won't forget it either.

Witness #8

(1) I was sitting outside talking with my friends, (2) heard a low buzzing noise and my friend pointed it out, (3) thought it was a UFO because my friends saw one and I guess it flew over them or something, (4) it was orbs they came together and formed a "V", then they continued to fly at speeds that were faster than a jet, made sharp turns and a yoyo motion up and down and flew east and left, (5) amazed we filmed it but I don't have a copy to give you, (6) it flew east.

Witness #9

Exited Walsh College Novi Campus at 10:05pm. Immediately noticed 2 bright lights due west that appeared too bright and too close to be stars. At first glance, one appears to be star-like and the other looked like a distant glimmering star. The glimmering star was glimmering far too rapidly, and one reminded me of the two separate experiences I had in Northern MI. So, I watched them for a while once I entered my vehicle. My vehicle was already facing west. Once stationary in my vehicle, I confirmed that they were both in motion, relative to one another, to the ground and my location.

I filmed and photographed with both my cell phone and an iPad. While filming on iPad, I could see other entities and lights in the sky that were invisible to the naked eye. The white sphere, star-like orb appeared as a green ball that moved in many directions, zipping up, down, diagonally, and sometimes sort of hovered. While in motion, the green ball moved very quickly with abrupt stops and starts. The glimmering entity also moved around, but it did not appear to make abrupt motions, and appeared much smaller and less bright than the green-white one. It rapidly flashed colors. They were so rapid that it was difficult to tell if there were more than 4: red, yellow, green, and blue.

There was also a very large entity that seemed to serve as the station or mother ship for the orbs. It was dark, like the night sky, and invisible to the naked eye. My iPad was picking up lights and other shiny, iridescent or metallic characteristics that appear to be large corners. The corners, themselves, appeared to be 3 stories by two stories in width. Overall, it is so large that it is difficult to estimate size. It was very low to the ground and very close.

In several pictures and video, a string of red lights are visible on the craft, but not visible to the naked eye. The back-end emitted bright orange flames a few times, and what appeared to be bluish gases. It appeared that other UFOs were going in and out of the large craft in the video clips on the iPad.

My reaction was calm. I was not immediately frightened. I have seen UFOs twice before and I wanted to be sure that I had more video and photos this time. I had to leave, but I wanted to stay longer. Before I left, I noticed a classmate outside smoking near her car. I pulled my car up and asked her to look up and see if she noticed anything unusual. She replied, "What? You mean the stars?" I said "yes, do they look different to you?" As she replied that they sort of did, I noticed her demeanor changed quickly, like she was frightened. I asked her if she saw that they were changing distance relative to one another. After she saw that she decided to leave immediately and offered to talk more about it at a nearby gas station. She was present at my passenger side window as I filmed a clip out my driver side window. After she left, I also left another minute later.

There were more sightings, but not so close or intense, from my home, near Lake St. Clair. All of them were different. I may make a separate report if I have time to do so.

Witness #10

I submitted this once already a few weeks after the event took place. I have decided to submit it again with the possibility of video evidence from a nearby public location with 24/7 surveillance which I have not been brave enough to ask for footage. They probably would not release it to me anyway.

On January 4th, 2011 which I believe was a Tuesday night, my uncle and I were driving from his home in Lititz to Molly's gas station in Manheim for a casual drive for coffee and cigarettes. It was app. 8:20 pm. It was me that noticed the orb first pulsating peach in color very high straight ahead as we reached the trailer park between the two towns. I was driving and chuckling I said "look it's a UFO!" The Lancaster airport is only about 10 or less miles back toward Lititz. So I see lights and planes and helicopters daily. At first it was strange, but it could have passed as a plane with its lights right at us.

We watched it closely and after heading around the bend at Doe Run Elementary School by the pond, this object fell from the sky. We thought it was actually crashing but *it did not. While at this point it is only a few hundred feet away, the lights are very bright! We were in disbelief because this object stopped falling right above the school building, hovered for a second, spun or maybe lights were just spinning horizontally then diagonally very super-fast! Then the object took off like spinning and was charging it for its extreme take off or something! It shot like a meteor but did not go straight. It slithered like a snake in the night sky but first made another sharp turn. I think it was bluish in color when it took off. It was completely silent. At that point when it took off I stepped on the gas. Obviously I wasn't going to catch up but I had to try! By the time we got to the light past the school, which is only 200 yards, this thing slithered past the west side of Manheim right over town and probably was by 283 - already out of our range of sight.*

It was a very clear night. I don't understand how we were the only ones who saw or reported this. There was a car back behind us but who knows if they were too old to notice it or what. My uncle and I don't talk about it too much lately because the conversation always leads to silence or deep thought.

To be honest this did affect me in some negative ways, but maybe that's because they just won't tell us the truth. I was in shock for sure when I came home and told my wife. In fact I was in shock for a good week. If your own wife doesn't believe you, who do you tell? It still worries me, but I am much more clear-minded and sane now. I just wish I could see it again...and again. My uncle believes it may have been something biblical, so I don't push the subject, but we both know it wasn't of our publically known technology.

Witness #11

The top description pretty much covers it, I have seen these several times last year and reported once on this site. I don't know what it is, nor does anyone that lives around here. They are basically very large orange orbs that turn on and off, and don't move, they seem to do something like a hyper-jump, if that makes sense - one turns off as another turns on, they don't fizzle out - they just shut off instantly as another lights up. I have seen this phenomenon while it was still rather light outside, and you can't see anything when they shut off - no object, no smoke trail, nothing. This event was brighter than I have seen in the past, so I thought I would report it and see if anyone else has seen it. My in-laws also saw the tail end of the event but they will be heading back to Hawaii, so I have not included them in the report; however, if it turns out to be necessary I am sure they would talk with you.

Witness #12

I was in my bedroom with my brother (we share a room) when I was looking out the window facing me. That's when I first noticed the red lights in the sky. At first I had thought it was a formation of military jets or some kind of helicopter, but then I noticed that two of the three balls of light were orbiting (circling) the third ball of light. I told my brother to look out the window and describe the object to me while I drew it down in a sketchbook. After a quick draw I looked up again just in time to see the object start to drop smaller orbs of light, I remember jokingly saying "it's like a Battle: Los Angeles invasion" but I wasn't quite sure what to think of it. I remember mostly feeling wonder, also a bit of fear when the objects began to fall from it and my brother may have felt the same thing, but we were both very quiet for the whole thing. We lost sight of the object about an hour later when it suddenly disappeared, maybe flying off or maybe teleporting I don't know.

Witness #13

I walked outside at about 2 AM or so to make sure my car was locked and I saw this weird blue glowing orb shooting across the sky towards a stationary white orb and a weird reddish glowing orb. And then if that wasn't weird enough, I watched the blue orb shoot off something that look like a beam or something at the red orb. Then the red orb kind of shot off for about 30 seconds and the white orb turned a reddish color all about the same time - and the blue and red orb started what looked like shooting lights at each other. After about one minute of this I watched the sky in the direction kind of flash like thunder for about 5 seconds and these things were moving in weird directions stopping and then changing direction. After the flashing, the red orb was gone and a second orb appeared. It was red also. I assume it was the first red orb that disappeared and they both shot off toward the NE, all of this sort of lasted I'd say 5 minutes. It was kind of like a Dogfight.

Witness #14

My friend and I were in route from Port Arthur to Houston, traveling west bound on Highway 73, we had just passed the landfill (between Winnie TX.) The time was about 11:45 PM. We were visiting my aunt, in ICU in Port Arthur. I had just activated (the Google Navigator) on my 3G phone. I was thinking to myself - how cool! We are tracking in virtual time on a satellite map..."this must be the way we appear from an aerial view."

At that same instant my friend who was driving, in a calm collective manner says, "See - I told you that there are UFOs around here!" I looked up, my eyes immediately fixated to the top left corner of the windshield. And ... sure enough!!! There IT was!!!! Traveling east bound, about 200 ft. above the highway. It appeared to be pulsating an iridescent/fluorescent bright rainbow color, however all of the colors displayed at the same time.

I find that words are very difficult to express completely what we experienced!!! There is just not any vocabulary that I know of that can replicate what we saw!! However, I will continue at the risk of sounding completely out of my mind...lol!!

There were three rhombus-shaped pulsating objects - they seemed to be separate, but yet as one unit?!?!? They were silently/fluidly traveling east bound, parallel to highway 73 ... I have always had a scientific curiosity ... I asked my friend "DID YOU SEE THAT?!?!WHAT THE HELL IS THAT!!!????" four times. I also rolled all of the car windows down and I opened the sun roof. This was to reassure myself that we weren't just seeing a reflection off of the car's glass and the fact that my friend confirmed four times that we were both seeing the same phenomena. As I opened the sun roof, never taking my eyes off of the object, I popped my head up through the sun roof, and said WOW!!!! That's no reflection!!! That's a bonafide UFO! That very instant, the object made an immediate U-turn!!!!!!!!! And now it was following us down Highway 73!!! I never once left visual contact with the object. I jumped down into my seat and shouted "THEY ARE COMING!!!!THEY ARE FOLLOWING US!!

Keep in mind that this is a dark and desolate highway at this time of the night. I kept my eye on it, while my friend was franticly speeding up to get away! I could see the same colors pulsating 100 feet above the pavement behind us, but it seemed very different than what colors you or I are used to???!!! I was almost hypnotized by this strange sight...That's when I realized that we were not only being followed but, also being monitored on a level that no human/animal could even begin to comprehend!!! I could see the triangle shaped pulsating rhombus shapes are getting larger, and closer, leaving and getting smaller, and yet not moving at all?!!?!?!?

Well my friend, as strange as it may sound, all of this was taking place simultaneously. My eyes and brain were not quite sure how to process all of this together! We became panicked and sped up to approximately 120mph. Again this thing never changed perspective, of course we were speeding, but the object stayed behind us, but not ever moving?!!??!!? My friend says to me "STOP STARING AT IT"!! The object got almost to the back of my car, then I noticed a yellow/amber-ish colored column of light slowly coming down to the ground from the object ... then the column of light became very, very bright, nontransparent, and became almost like an orange fire ball. I could see the grass and the dirt flying about, under this fireball!! I started screaming" "THEY ARE GOING TO BLOW US UP!!!!!!!" That instant, the object started to wobble and the air all around it seemed to ripple as if it were H20?!?!?! Then the object became one with the column of orange/yellow light. It reminded me of how Mercury looks on a hard surface. And all in the same split second, the object then became a bright large sphere, with the same pulsating colorful pattern ... I was shell shocked!!!! Did I just see that take place in this dimension/this planet/this whatever!!!!!

By this time I could feel a presence in my persona...not anything unusual...no strange voices, but almost like my subconscious was suddenly in charge?!!?!I never once felt threatened in that sense, but I was subtlety able to become aware of this new/unfamiliar feeling in my head?!?! Sounds looney, I know!!!! But like I said, there is no vocabulary that we have as a species that could come close to properly describing these events, senses, feelings, this "phenomena".

We continued for a few minutes, then I noticed two cars traveling east bound were coming after what seemed like an eternity!! I was wide eyed of course waiting to see if those cars would notice what was following us. Sure enough ... they all slammed on their brakes and pulled to the side of the road!!! At that time the object behind us, backed *down and in a split second, the sphere divided into two and disappeared!!!!!!*

I feel like someone else out there had to have seen something. If so please share. I feel a sense of isolation from people who have never experienced these phenomena. It's almost like I cannot communicate to them the events that took place that cold clear starry night! I can't help but look to the sky every moment I am outdoors!!!!

I am still a bit "rattled" - not threatened per se, but I have a sense that who/whatever that was, is still out there, and knows where I am and what I think before I think it!!!!! Well - there you go!!! That's my first encounter/interaction with anything/one that has that magnitude of capabilities.

Witness #15

I had just shown up home from running down to the convenient store to grab some things. When I entered my backyard to get inside, an orange orb came up from behind the trees. (Ascended from 10ft off the ground to 30 ft. in air. around 50 ft. away flying out from behind the trees like it was waiting.) It's gaining altitude, moving in an aerial manner, almost arch-like, at me. The first several seconds I look at it, it's amazing, immediately knowing this was unlike anything I had ever seen before. Then I realize its gaining speed and descending straight towards my direction. I became extremely panicked and bolted inside at a "Run for your life" speed. Upon entering my home, my family was playing around with each other, obviously oblivious to what I had just experienced. But I felt so shaken up. I just tried to say nothing and alarm no one because I felt as if anyone looked outside with curiosity something very bad was going to happen.

After about 10 minutes inside I'm telling myself "it's gone" why would it stick around? I felt like the orb had left and enough time went by that an object moving at that speed would be at least several miles away by now. (Roughly 15 minutes later) Me and my girlfriend at the time, went downstairs to my room. I couldn't handle not telling her so I did. She, without a second thought, ran over to the window. I'm talking her away from it. "Babe please! Come on - it's gone, etc." but she was already looking out the window. I stood on the far side of the room saying "Katie? Katie, what's up?" I walked over and peered out behind her shoulder and there it was - the orb. It was hovering on the other side of the lake, 400 ft. or less. Almost the instant it saw me, it shot towards me at a speed I have never seen.

I have no memories after that besides a mental snapshot of jumping or something into my bathroom with both feet in the air as though I was literally jumping away from the orb. Then we woke up the next day as though nothing had ever happened. Over the next month or so I would steadily regain memory of that event, although that block of missing time remains just that, and I have NO memory of anything after it shooting towards me.

Witness #16

I am an amateur astronomer with a 13 inch diameter telescope. I have been viewing through this scope for about 15 years. I am a former pilot too. I set up the scope in my backyard in Nipomo California. It was dusk at 6:35 pm. I pointed my scope at what I thought was Jupiter. (Jupiter turned out later was hidden from my view at the time by a tree). This event lasted at least 3 minutes. I looked at the object - it had roughly the same diameter as Jupiter through my scope using an eyepiece that gives me 49 power.

As I looked at the object which I thought was Jupiter, there was what I thought was one of the moons of Jupiter, the smaller object, which started moving back and forth above the larger object in about a 180 degree arc lasting about 1 second. I watched this go on for at least a minute. I checked my scope and thought perhaps I was picking up a reflection. My optics were ok.

As I looked at the two objects, the smaller one appeared to shoot out a small beam of light - a laser? The color of the beam appeared to be a combination of green and orange. The smaller object continued to move back and forth in an arc above the larger object while aiming the beam. Both objects stayed in my field of view - they did not move out of my field of view or were stationary in my field of view other than the smaller object shooting the beam moving back and forth above the larger object.

I watched this beam shoot out of the larger object for close to another minute, then the larger object exploded into at least 2 dozen pieces - each a bright white like the original larger object. I then watched the object that shot the beam move slowly away from the debris cloud. I tracked it with my scope. It then appeared to take on an elongated shape of a very dark red rust color. It also had a flashing light at the rear.

I moved my scope back to the debris cloud and back to the object that shot the beam a couple of times. Then the debris cloud was gone and I lost the object that shot the beam. It was later after the incident was over that I realized that Jupiter was still behind a tree in my backyard. The object when I saw it was at the same elevation above the horizon, but about 15 to 20 degrees north of Jupiter. The object was almost in the constellation Pegasus. That is all I have. I was the only one who saw it. I do not know what I saw - I cannot explain it

Witness #17

First, this happens a lot, but these particular days (Saturday and Sunday night), I was able to see more activity and make out that these aren't just reflection lights or lake lamps, etc. The lights come as soon as darkness falls, but don't seem to start moving around until later - around 10 or so. I've seen the same thing repeatedly, I've already reported what I saw earlier in the month.

Basically, a bright light will be hiding (not too well) in trees in property across the street and then dimmer spheres dance around in the spotlight that the brighter light provides. At first I thought they were animals, but I can't clearly make that out. What I <u>*can*</u> *clearly make out are dim spheres that seem as though they are looking for* something - they go in and out of focus, so I think they could be changing shape, but I can clarify they are spheres for periods of time. Then lights that change color fly across the lake - I hesitated to report this because it could be jet-ski lights or whatnot. But Sunday night I saw clearly - these are just spheres of light and I was able to see the reflection in the water and they were so bright - no jet-ski for sure.

I don't think I'm the only one noticing - we just moved here so I don't want to go ask the neighbors about the "UFOs", but I noticed a boat came out several times turning off its lights and checking things out. Then last night I saw the sphere go towards my neighbor's boat house and then disappear...next thing I know the neighbor is out with a flashlight looking for something around his boathouse.

Also, way up in the sky - like a star - it seems this object dances around from place to place and blinks and dims, seeming to communicate with the other lights. The ones lower to the ground don't always stay together - they will split up and the star up in the sky will jump from place to place.

Witness #18

Around 1am I was called to the window to witness what were at least 100 star-like crafts dropping into the Kill van Kull. No surprise someone from California just put up a report of exactly what I saw. I and my girlfriend were up till 4:30 am watching these low dropping objects. This is not the 1st time I have seen crafts above the Kill van Kull. Out of all the crafts, the most memorable was a fiery disc that shot across the sky in a strange wobbling motion. Some crafts were also not lit up, but you could see them with the moon light reflection, dancing and playing with each other like kids. Very peaceful display in their aerial acrobatics. Since I was a child I have seen crafts of all kinds. Last night was something surreal.

Witness #19

My friend and I were out in a farmer's field with our packs in North Dakota getting ready and in shape for our climb of Mt. Borah. We stopped to drink some water when my friend noticed four objects next to the unusually large moon. When further looking at the round bluish-white orbs, they started to move in a circular formation, keeping the original diamond pattern. After watching for a limited time they slowly formed a line and headed north northeast.

While watching, they made drastic zig-zag patterns towards us until they were right on top of us - then almost slowed to a stop. At this time I was fumbling with my phone to video all this but my phone was frozen and unresponsive, so I removed the battery from my phone. After removing my phone, the balls of light left slowly but one broke off and disappeared. We watched the three disappear in the sky and decided to continue on our light jog.

We made it about another mile in the direction of where we first saw the lens. All of a sudden there was a bright light flashing in the trees the size of a flashlight with no beam that exited the trees. We shined our light on it and saw a three foot figure. Then we turned off our lights and the flashing started again in a different spot. This happened six different times. We decided to turn back and head home. The fields here are broken up by tree lines. We crossed four tree lines and in every tree line behind is the small entity and the glazing light at us. We finally made it to the last tree line. After we were about a hundred yards from the tree line we stopped and looked back to see the small light still flashing at us.

Then all of a sudden, there was a bright strobe light that rose above the tree line and moved along the trees following us to the highway. By then we were very frightened and started at a fast pace to the highway where there was traffic on the road. We made it to the highway and the object stopped close to the highway and kept strobing. It was a very bright blue/white light. We decided to keep moving to our camper. The light disappeared so we thought we were in the clear of getting taken until we reached a nearby radio tower and saw the small light in the trees again and a very strange hovering noise.

We were once again freaked out and started a fast pace again towards the camper a half mile away. We reached our camper and saw no more lights but could hear stuff moving in the grass twenty feet in front of us but would stop when we turned on our lights. We decided to go into our camper and wait it out until the morning. All of this took around two hours from the beginning of the sighting to the time we got back to our camper. We walked roughly three miles to the sighting which took two hours

Witness #20

My year old aunt and I were at a local sports pub last night out on their back deck having a smoke and seated at a table where two other people were also sitting and smoking. My boyfriend looked up and said, "What the heck is that up there?" And all four of us looked up and there was an enormous glowing red-orange sphere moving slowly and silently in the sky. It was simply enormous, about softball-sized and it moved very, very slowly, changed direction, then paused and just hung in the sky. Then it proceeded across the sky again slowly and when it was almost out of our sight, it broke into two tiny orange pinpoints of light and each pin point zipped fast away in opposite directions and were both gone almost instantly.

We discussed how large we all thought it was and the consensus of everyone who saw it on the deck was that it was very, very high, and extremely large and we have all seen satellites moving on a clear night, and this was moving way, way slower than satellites and even slower than a plane. It was too large and silent to be a helicopter and none of us could discern a shape - it was more like a glowing red-orange sphere and none of us have ever seen anything like it. I decided to report this because I went on your site and saw that there were a number of glowing orange spheres last night recently being reported in California and New York, so I thought what we saw was so much like what those other reports described that I wanted to add our sighting to your site.

All of us discussed that it definitely didn't have the feel of anything man-made or terrestrial, and it gave us all a really weird feeling in the pit of our stomachs and the thing that was truly amazing to me is that it broke up into such smaller pinpoints of light...how that huge sphere that was so darn bright could break down instantly into two smaller way less intense lights was pretty incredible and we were all pretty stumped and amazed at the same time.

The entire sighting probably only lasted 30 seconds and we were so stunned that unfortunately no one had time to get out our phones to snap a picture. It was so dark out, it would have just probably been an orange blob against a dark background.

Witness #21

As reported to ASD by phone and e-mail: I let the dog out of the back door of my house because he was barking at something — at first I thought he just needed to go out, but he stopped and looked up after he reached the edge of the patio. I stepped outside and looked where the dog was looking and saw three spheres floating at about 30 feet off the ground not 50 feet away from me. They were glowing a bright red orange and looked like they were on fire.

My first thought was that they were meteors, but I realized that they couldn't be. The objects were close together — maybe 10 feet apart — and were probably each three feet in diameter. They moved slightly while floating there and kind of pulsed. I was just too shocked to move and afraid to draw attention to myself. I did not have a camera so couldn't get a picture. They stayed in that position for approximately 5 minutes, then moved slowly off in an Easterly direction — as if they didn't care if anyone saw them or not. I did not see wings, or a tail or any lights.

The next day I spoke to some neighbors about it and one said they saw something similar earlier that evening but that it was only one ball of light, not three. He said it was right at tree-top level and then shot off quickly to the East. I just thought that someone should know about this since there have been so many reports lately.

Witness #22

I am fascinated by space and the stars, so when I am outside at night I always look up. Well as I lit my cigarette in the backyard of my house in Clovis, and began to gaze up, I noticed a star-like object in the sky as I looked to the southeast. It was flashing many different colors (red, green, blue, purple, and white). Not only was it flashing colors, but it was pulsing, dimmer, brighter, dimmer, really bright, dimmer etc.

At this point I knew I was definitely not looking at a star or any type of airplane or helicopter. Then I saw it do something I will never forget - it moved! It didn't just move - it zig- zagged down, and then shot up at a fairly rapid pace. I don't know of any man-made aircraft which can hover, stop, and instantly change direction. It also could go from a very high rate of speed to a complete standstill.

When I got over the initial shock, I ran inside and woke up my roommate. I was so excited I yelled and shook him rather hard. When he woke up he was terrified and upset that I woke him up so abruptly when he had school early in the morning. After he heard what I was yelling at him we both ran back out in the backyard. He even noticed the object moving and changing color without his glasses and turned right back around to run in and get them and a pair of binoculars. I stayed to keep an eye on it and he came running back out and was even more shocked when he saw it with his glasses.

We both took turns using the binoculars for about 45 minutes and that's when we spotted the second one. It was more to the east than the first one we noticed. It was doing the exact same thing; pulsing, changing color, and making very erratic movements. However, this object seemed to be higher up. At about this time I had my roommate go get the Fresno Air Terminal Security phone number. I called and spoke with a young man who said he couldn't get outside to see but he would phone the tower. The two officers in the tower did admit they had seen some strange blue/green lights, but they did not think they were threatening so they didn't investigate further.

My roommate had to go to bed because he had school so early. I continued to watch as I proceeded to call anyone and every one of my friends and family to try and wake them up so they could see. Unfortunately, nobody woke up and my webcam wasn't good enough to see anything on. I continued to watch for another hour then called Fox 26 newsroom at 3:45 or 4am and spoke with a young lady. I was hoping they would have a camera handy, they did not. (I found that hilarious)

About this time I noticed something even more shocking, I could see many more that were farther away (not higher). In total, I counted at least 9 or 10 different objects. The next day I contacted MUFON and learned a local police officer had reported seeing the same thing! I felt so validated when I heard that, and that made up for all the crazy looks I'm going to get telling this story.

This was the most incredible night of my life, and I will definitely never ever forget it. I know some people I talked to last night and today might not have completely believed me, but both my roommate and I know what we saw and we will tell anyone who will listen! We have a video camera tonight so bring it on! (I am currently working on a sketch)

CHAPTER THREE
WHAT'S INSIDE THE ORANGE CLOAK?

Witness #1

I am a bus operator for San Diego Transit and was driving down La Jolla Blvd. northbound at approximately 7:40. I am believer in UFOs and constantly look to the skies while driving the bus at night. I see airplanes and helicopters all the time while driving. What's interesting is that I always look for the orange orb as opposed to the bright airplane lights.

As I'm driving I observe what appears to be an orange glow in the sky. La Jolla Blvd has roundabout circles in this area, so as I look up and make the turns the glow comes in and out of view because of trees and buildings. I then think to myself that it's one of those Chinese Lanterns until I realized it was coming towards me as opposed to floating. This is out of character for me, but against company policy I pulled the bus over at a bus stop located at Bird Rock Avenue even though there was no one at the stop. I stepped off the bus and took off my hat and watched the glow. It looked like the orange orb that I have heard so many UFO enthusiasts talk about.

As I watched it travel south it seemed to disappear and re-appear as the glow pulsated out and then back. I then looked north and saw a plane that appeared smaller than the orb. Seconds later it went out again, but this is when I observed the object. Even though the glow went out, I could still see a round grayish object moving at a high rate of speed southbound. I watched it as it flew and realized I needed to get back on the bus and did not wait for it to vanish from my sight.

Witness #2

On Saturday November 13th around 6:58 pm eastern standard time, I was on the way home from a local mall with my son and my nephew. I noticed something out of the corner of my eye that was glowing bright orange and this caught my attention immediately and I pulled over my car and investigated further. The glowing orb began to drop several white looking balls of light out of it toward the ground. I knew this was not a plane so I took out my cell phone and started taking pictures. I also made an eleven second video.

There were two objects that showed up on the video. The one object hovered on the horizon and as a commercial airplane approached in the same direction as the object, the lights dimmed down and there was a flash of light and it was gone. I came home and reviewed the pictures I had taken only to be shocked by what showed up on film and video. There were multiple objects on film and they appeared in different shapes, colors, and sizes. I also have pictures that appear to show the face of something standing in the wooded area in front of us. It was not a reflection as there was nothing but woods in front of us. The video I recorded showed two orange orbs - one moving and zig-zagging off my screen.

I don't know what it was that I saw. But I would like for someone with better equipment to view my video and photos and let me know their thoughts. I have more pictures to submit. Also last night November 23 I took some more pictures of a different object in the same area. More to come as I will be uploading those pictures and new video as well. Thanks

Witness #3

I was driving to Waupaca WI and was a little lost when I observed a fireball in the night sky. It came through the clouds. I thought it was a burning airplane falling from the sky until it turned and headed my way, leaving a trail of red smoke in the dark black night sky. I woke up my wife who was sleeping in the passenger seat.

The object traveled about 3 miles to my location and turned into a metallic cylinder with fire emitting from holes along the side. It was completely silent and I saw beings in the windows. I pulled over on the shoulder of the road and turned off my engine. I looked at it and prayed. I did not reach for my weapon or my camera because I did not want to put my family at risk. This ordeal lasted for about 15 minutes. As I began to wake up my son who was sleeping in the back seat, the object turned into a bright white straight line pointing in a downward direction. I then noticed a pulsating white light on the ground from across the road. I think it landed and I got nervous and started my car and drove away.

I told my friends who we were visiting and I called the Sheriff's Department who told me they had no UFO reports for that night. I am a former police officer and have practiced law enforcement for 30 years. I have told few people about my encounter and have had unusual dreams over the years.

I made a promise to the UFO at that time that I would not go public with what I saw if they would protect me from evil and bless us with success. Before we left Waupaca WI we went back to the area and discovered that the object descended into a stream.

<u>Witness #4</u>

The object moved very slowly over the tops of the trees, and eventually I could see it entirely. First off, this event happened on the night of July 4th, 2010. It's easy to suspect that I and four friends misidentified some sort of fireworks display, and that's what I thought it was until I really got a look at it, but I will simply start from the beginning.

I and six friends went out into a field to shoot off some fireworks to celebrate Independence Day and we had been out there for about two hours already when the event occurred. We had already used up our supply of fireworks and we were discussing what to do next. Two of my friends decided to go into town to purchase more fireworks, so that left five of us behind in the field waiting for their return. We weren't really talking to each other, we were pretty much all just enjoying the weather and looking up at the stars. The moon was not visible that night.

Well, I noticed the object before anybody else did. I could see a fairly large object glowing through the trees to the southeast. The object was a mildly copper-colored (the actual color is difficult for me to describe, like a mixture of gold, silver, and copper colors) sphere which had flames erupting from the bottom of it, and rising up around the bottom of the sphere before finally dissipating around halfway up the sides. It emitted an orange-ish aura, not unlike the color of flames. It pulsated and flickered, much like light cast by a candle. As it slowly drifted westward, it didn't seem to change speed, direction, or altitude at all. It took over five minutes for the object to pass completely over the field and disappear behind the trees on the opposite side from where it had originated. I could clearly see the spherical shape of the object on the top, and I could see the flames rising around the sides of it. The flames seemed to me to originate somewhere on the bottom of the object.

The relative size of the object from my vantage point was anywhere between the size of a dime and a penny, but it was just around that size. Without any useful points of reference, I can't say for sure how high it was, but I would estimate an altitude of at least 200-300 feet, as it very easily passed over the trees without casting any light on them at all. It was moving at approximately 5-10 miles per hour almost directly eastward. It didn't make any noise at all, and the object didn't change at all during its visible flight over the field.

So far as I can tell, it was no firework in my opinion. All of the fireworks that I have ever seen simply can't burn that ferociously for that long while staying airborne. I estimate the object to have been far too large to be part of any type of firework display. At least one person that night tried to record a video of the object with their cell phone, but the low-quality camera couldn't register the object at all on the viewfinder. I can have more information readily available if requested.

CHAPTER FOUR
ORB COMPANIONS

Witness #1

July 5th, 2012 around 10:15 pm. My wife and I were outside watching stars in the Northern Sky behind our house which is a cornfield - very low corn with complete open view of horizon and sky. We both noticed at the same time, a very bright red light appear in the sky about a mile or so out to the northwest. The light hovered for several minutes before changing colors to a whitish-yellow light. It proceeded to move left and right but hovered in its general area.

I ran in the house to grab my camcorder which unfortunately was NOT charged up. I ran back out in hopes of seeing the object (my wife was still out watching it. It was still in the air, but appeared to be moving closer to us in a southward direction. Several beams of light shot from it, and at one point an object seemed to go INTO the object. The object appeared to change into a "mechanical" look and as it flew directly overhead our house we saw that it was a very large triangular craft with a light at each corner and with one very bright white light in the center. Another thing we noticed were several searchlights came on off the horizon from the north and the northwest. The object flew directly over our house and headed to the south where we lost sight of it.

A few minutes after the object passed, a dark double rotor helicopter with a searchlight flew in the area the object came from. The helicopter flew over the cornfield for a few minutes and then left the area. The searchlights also went off when the copter left the area. I've never seen these searchlights before. The object was apparently to us very large - we guess larger than a jumbo jet.

Witness #2

I'd been seeing the UFOs for the past few months now. I had a feeling last night to look. Usually they come around midnight, but last night it was 8 pm and I'd never seen them that low before. There were several of them in the sky but two were closest to the ground. They were hovering and were very bright. They looked like circles with flashing lights in all different colors of the rainbow. They stayed for a few hours before raising into the sky, they were quiet. The other was cigar-shaped and it was yellow, I'd never seen that one before. And I recorded them on video camera. I plan to continue documenting them. Since this happens a lot now, I've noticed a pattern. They leave when the helicopters come and the next day the helicopters stay circling in the same area the uses were that night.

Witness #3

So, I had reported a UFO sighting for my sister about 3 weeks ago. I hadn't seen it myself. It was a daytime UFO & she was very convinced in what she saw. So, naturally this piqued my interest, and decided to take it upon myself to do a little UFO hunting of my own. Almost every night since then (and even during the day) I would check the skies, standing out there from anywhere between an hour to two hours.

I saw a few things that I thought were 'strange' but was still on the fence about it. Until last night. I was at my sister's and we decided to go outside and see if we could see anything. Almost immediately after we started looking at the skies, it began. I remember standing there and exclaimed, "OMG! Look over there! Wait, look over there! Wait! Look over there! THEY ARE EVERYWHERE!" Sounds comical, I know. If you were there watching us and not the sky, you'd be laughing at us. But if you were watching the sky, you'd have been freaking out too! At first it was just the flashing lights all over. Glittering the sky. Very similar to planes, but these were no planes. Many of them hovered in one spot. Some of them moved around slowly in strange patterns. But they were popping up everywhere. They were all over the place. You'd see one and then you'd see a dozen more popping up all over the sky. And then there were these strange twinkling stars. Or at least we thought they were stars. Until they, too, started moving. And around them you could see 'shooting stars'.

If all that wasn't incredible enough, what we were about to see was mind-boggling. There was a glowing mist in the distance. It wasn't too far from us. If we ran for only a few minutes, we could have reached it. At first we thought maybe it was just a fog and the light was catching it in a strange way. Until we started seeing HUGE translucent things flying in it...several of them – two or three. They were enormous. They were close to the ground. They were only partially visible. They looked like ghosts in the sky. That's really the only way to describe it. It was the freakiest thing I'd ever seen in my life. My sister and I were screaming in terror and excitement. They did their circling dance for about 15-20 minutes. It was almost as if they were putting on a show just for us. It certainly felt that way. And then one of them must have shot up way high in the sky. It was no longer a shadowy/translucent UFO. It was saucer-shaped, glowing, and very solid-looking. And it darted to the left, southward. Like it was going to land! And almost before it reached the ground (which had to have been over the town park), it just vanished. So weird. Naturally, we were really freaking out and couldn't stand anymore. Too excited. Too overwhelmed. I mean, it was life changing.

I'd never seen anything like this before. I'd always believed. And I'd watched you-tube videos, but until you see something like this with your own two eyes ... it's not the same. But things didn't end then. We were sitting down and then we saw an unblinking set of lights - five lights. It was no plane. Like I said, unblinking lights. It was like triangular/pyramid shaped. It was so hard to tell, because it almost seemed as if it were shape-shifting. And it just flew across the sky, slowly, spinning. Sometimes hovering. Floating. These sightings lasted for about 2-3 hours before they started to slow down. It was amazing. If you're a skeptic, my advice is go outside. Know that you're going to see something. Because you will.

Witness #4

On the evening of December 29, 2008 I was at my brother's apartment for the weekend along with my girlfriend. I had the first sighting about 8:00 pm when I thought a police helicopter was coming over someone's house we were parked in front of. My girlfriend was inside, my feet were propped up on the dash of the truck, and I saw the lights coming over the house. I got out the truck to get a better look, there was no noise from this light. I noticed there were two of them. Suddenly they zoomed close as if going 1,200 mph and froze within several hundred feet ahead, and about one hundred feet above me. They had spot lights, windows, and were bright. We got anxious, and left after my girlfriend saw them.

Later on that night at my brothers, I began looking out the window, and found those glowing colorful lights as in the account recently posted on this site. We watched lights all night long. There were sparkles in the distance, there were red amber contrails passing right ahead of the house as if a huge glowing comet flew by, only inside the head it was shaped like a disk, but nothing there as if invisible, then it was gone in a blink.

We got in the car, and went to the river where it was very dark, away from the city lights. We watched many bright lights move, they look almost like stars. One especially we watched bob up and down, and all around behind the trees. We decided to leave, and the light followed us home.

At this point we began flashing our car lights at the object and it responded with bright colors. After about 6:00am we saw a saucer object pass directly over our car. It had no lights on and disappeared in the distance, it made no noise.

Another object, saucer-shaped, passed over the house at about 85 feet up in the sky. We hung out the dormer window and watched delta shaped lights blink from the back to front in repetition, it made no sound, and sped away very fast. There were lights everywhere that night, things that we three will never be able to fully understand, or explain. The last sighting was at about 7:30am.

We were all upstairs, my brother peeking out of the dormer window looking for anything else. The sun was up by this time. He looked left (due North) and said, "What the hell is that?" I looked out and saw a saucer wobbling, and hovering above the trees.

We told my girlfriend to get up and look, and when she hung her head out and saw it, she closed her eyes and looked straight ahead stating that she didn't want to see anymore, she had seen enough, she was scared, and in disbelief.

My brother and I continued to watch it though. It was silver, about 80 feet across, had two lights blinking on the top, orange, and red. The saucer wobbled as it moved, the outer part spinning as the center part was stationary. It moved up and down over the trees, and didn't make a noise, it was approximately 300 yards away. The whole time it moved we heard shotguns going off, people were shooting at it, as hunting season is still in right now in Statesboro. I estimate we heard gunshots non-stop for four whole minutes. They were either scared, or upset people, and it sounded like a war zone.

It finally vanished, bobbing up and down over the trees, at very low altitude, we don't know where it went from there. The gunshots stopped and no one saw it again after it hit an open area, dipped down, and shot up at a high rate of speed is my guess. It was dead quiet after all this passed.

We saw some triangle shaped objects the next night and some other strange lights, but nothing like Monday night and Tuesday morning. I've always believed in UFOs and the fact we are not alone, but my girlfriend didn't. She believes now, and doesn't look at me crazy anymore. I'll never forget that night. We keep our eyes to the stars now, they might not really be stars. I can't fully explain the sights, and emotions we all felt, to really know, to see, the questions left in your head, feeling like you're crazy. We weren't drinking or high. You almost want to suppress it in your mind, it's too much all at once.

Witness #5

The major sighting was on April 16th, but I saw it first on April 9th at 9:40 pm CST. I was checking on my horses, on our 3 1/2 acres in Collins, Mo., because one is close to foaling. What made me notice this was a BRIGHT RED Light that caught my attention at tree level, just north east of our property. There are frequent helicopters flying over our house because we are in line for the life flight from one small hospital to the bigger ones in Springfield, Mo, but this was NOT a helicopter.

I tried to dismiss it as an airplane, but there was no noise and it was as bright as the cell tower lights SOUTH of our property. Then, a few seconds later, 4 or 5 BRIGHT ORANGE lights started lighting up from left to right (about 1 second each) and going out about the time the third light in a row lit. That was a little north of the red light (as though it was moving VERY SLOWLY to the north). Then the orange lights lit up again, a little farther north but not very much, (if I had to guess how far they were from the first red light I'd say they moved about 10 feet over a period of 5 seconds with each lighting). This time there were 5 or 6 BRIGHT ORANGE lights. These went left to right too, only the last couple, sort of went around the object, as though it was a circular object. These orange lights were, I thought, sort of computer-looking and more rectangular than the ones on the 16th. I did see them from a different angle than the ones my husband and I saw on the night of the 16th. If they moved north, as they seemed to with each string, they MOVED VERY SLOWLY. These lights did the same thing that the first set of orange ones did - they went out about the time that the third light was coming on.

I ran to the house to get my cell phone, hoping to take a picture of it, looking back as I ran to see if it was still there, and it was not. I check my horses every 30 min to 1 hr., and kept looking but it did not come back. I was alone that night, because my husband is a truck driver. I was excited to see it, but a little frightened because I was alone. On April 16th though, he was home and he saw the lights with me. We got our video camera out just in time to catch the last couple of times it lit up, but couldn't get a very good shot at it. My husband was filming in night vision on the first light that he caught, (which were the last two strings that we saw) and didn't get the focus very well. The light showed up as white, on the night vision setting. The next shot he got was on regular vision and he focused in on the last two of a string of four of the orange lights just before they went out.

The sighting of April 16th was virtually the same thing as I had seen the week before, red light started and went out, then the orange lights began lighting up left to right, then one time they lit up right to left. Also the lights looked more like orange fireball shaped flares but they didn't go up or down or anywhere but out. Again they lit up left to right, about one second apart at first, but, on one string of them they lit up a little faster and went in the direction like it was going around an object, and the last few lights in that string were closer together (my husband said it was like the lights were forming a sort of arrow).

This night, April 16th, the sighting was longer, with more light strings (that's what they seemed to be doing, stringing left to right, with a little curve at the end), and some differences in the color of the lights. The first ones were in the north east again, they didn't seem to be moving north like the ones on the night of the 9th, there were different amounts of lights in each string, one having three orange lights, and one having one, and one having 9 or 10 (the one that my husband thought looked like an arrow, it pointed ESE) almost like a code. Then we saw a little later, more yellowish orange lights with more white forming a string of three lights straight east of us (these ones are the ones that showed up white in the night vision shot). Then 4 more orange lights more in the north east again (these are the ones that he got a shot of the last two lighting up, (As though there were two UFOs signaling to each other really close to each other or one very big one with lights going in the south end and the north end at different times) NOTHING MADE ANY NOISE! We could not see the source of these lights at all, BUT, it blocked out the lights that usually silhouette the tree line east of our house. That would be about 1/2 to 1/4 of a mile span. We could not see any stars in that area, and usually we can see a few dim ones. The source of these lights would have had to be black, big and flat, if the lights were coming from a craft of some sort.

Then my husband saw a red light in the timber behind our neighbor's house which is east of us (closer to the ground). I saw a white light at the time he saw the red one closer to the ground in the timber. Then the last thing I saw, which no one else saw, was an *orange fireball shoot right over the roof of our neighbor's house (east of us) from SSW to ENE. We had put the video camera on the charger by that time and my father and mother-in-law had come to watch the light show. They got there too late to see anything except what we had taped. This sighting lasted about an hour. Then it just stopped, no noise, no lights nothing.*

I did hear though the next day, that someone who lives east and south of us several miles also saw weird lights in the sky about the same time that we saw these. Neither one of us has ever seen anything quite like this before. I have seen the big black triangle thing with the round lights on each of the three corners while driving a truck at night, and I have seen a very long silver "ROD" shaped cylinder over my father and mother-in-law's house during the day several years ago, among other things. But nothing that anyone else ever saw at the same time that I did, so I never reported them. This time, there was another witness, and we got it on video.

I don't know how to attach video clips or sketches to this report, but I did call the St. Clair county sheriff's department and reported the sighting of April 16th and what I thought was the previous sighting of the "Thursday" night ones that I saw alone. I keep a dated and timed hour by hour diary of my horse's progress through the last stages of her pregnancy, so I had written what I saw in that diary. After giving copies of my notes of those lights sighted on April 9th (still thinking it was the previous Thursday), I realized that the date that I jotted it down in the horse diary was April 9th. One week to the day before the April 16th sighting. The deputy didn't take the tape because he said that there wasn't enough on the tape to show anything. I told him that this wasn't a joke, after he said he didn't know what he was going to do with a report like this, and declining to take the video tape.

I'll never report things like this to the sheriff's office again in St. Clair County. I told him that I wasn't a kook, that I was an RN. And that I knew the sheriff, and the officer in the next small town north of us, and that they knew me personally, but that didn't seem to matter to him! I wish I hadn't called it in now!

<u>Witness #6</u>

My friend and I had decided to take a quick cruise on our motorcycles at about 10:15 pm. At approximately 10:30 we were coming up on the final turn onto our road. My friend was ahead of me. He made the turn, and as I came up to it, I thought I saw an orange light out of the corner of my eye directly north of us. But, when I looked up, nothing was there. I looked back down, and saw my friend had shut his bike off in the middle of the street.

I pulled up and asked him what he was doing. He said, "Did you see that?" I said, "The orange light? Yeah, I saw it for a split second, but it was gone." He said that when he turned on our road he could see it for about a full second, and it simply disappeared. We made our way back home, and stood around in the driveway talking about what we had seen. About 5 minutes later (10:36 pm or so) we saw the light again. It was low enough to get lost behind some trees. It was about as large as if you held a pea out at arm's length, and very bright. We ran out to the road to re-acquire the light. We watched it for about 30 seconds, it was north of us, heading south, and right at us. It didn't seem to be traveling very fast. Then, like before, it disappeared. It didn't look like something was simply 'turned off.' It looked as if the light caved in on itself.

We went inside to grab the video camera, sure that the light would return. We went back outside, and waited. We didn't see the orange light again, but, at about 10:45 pm, to our disbelief, we saw what appeared to be a triangle with a solid red light at each corner. There was no sound, and no flashing lights. It was traveling SE at a pretty high speed. Assuming its altitude to be around 30,000 ft. I would say it was moving at about 500-600 mph, and would be much larger than an airliner. At this point, we realized the battery for our camera was dead. 30 seconds later, it was only visible as a small red speck in the sky that faded to nothing.

A few minutes later my friend was in the garage and I was standing just outside of it. I glanced over our house, and saw what I thought to be a very bright star. I looked away, thinking to myself I didn't remember seeing a star that bright in that position in the sky. I looked back and noticed it was moving north, slowly. As it moved it was dimming. By the time I yelled at my friend, and he ran out, it was about half as bright as it was originally. We watched it slowly move north and fade in intensity for about 20 seconds. Before it was totally faded out, we observed a 'core', something solid the light was coming from. It simply was not visible once the light had dimmed completely.

We were scared through each of these observations. I am unsure of what, but, there is definitely something going on out here. This is not the first time that I have seen strange things in the sky in recent months. I have seen things by myself, as well with one or two witnesses (my roommates).

Witness #7

Our dog barked to alert us of trouble, as he had done with a past sighting. I went outside and saw 3 objects with bright, flashing lights. One stayed stationary as 2 moved away. One of them went into a stealth mode after a circle of red flashing lights. The other followed suit and both appeared again, only more started showing up from the area of the stationary cylinder shaped object. One of the later ones appeared to be cigar-shaped as was spotted last year. Several shapes and colors of lights were observed. They mostly headed in a SE to SW direction, some high up and others low to the ground in appearance. Some seemed to try to camouflage themselves, like they were just heading to the local airport, but they were way too large and anybody watching could tell the difference.

I had been watching TV with the family when our dog alerted us. Other family members saw this as well. The airplanes have an audible engine noise and these guys were only audible to the animals and their extra senses. This is another dead giveaway that they are not of the norm.

The way the objects jump around at times and their size also says this is not right. Either our government is up to something with unknown aircraft, or the dark angels are revving up their appearances. I first saw them just before 9pm CST and saw last one so far around 10:05pm. Honestly, I have seen many here in the past, but nothing like tonight. Wake up people and be ready, for the time of the Rapture draws nigh.

CHAPTER FIVE
TOUCHING DOWN

Witness #1

My girlfriend, best friend and I were camping. It was about 2 o'clock in the morning. I took a walk away from the fire and when I looked to the sky by a lake we have and saw a round object flashing multiple colors, I called the people with me to look at it. After this we stayed watching it.

It was so amazing to see how it moved. It was moving in a circular way, but it was emitting different light beams from it. It almost looked as if it was scanning the surface of the woods. I got an idea to shine the light at it and when I did, it reacted to it. I would shine it to different points in the air and it would fallow the light I was shining almost in the exact time I would change direction. This is when it seemed to get closer and make even crazier maneuvers in the sky. When it started getting closer, another one appeared behind it a little ways, but this one seemed smaller than the first. They were keeping a good distance away from each other and I decided to shine the flashlight directly up from where we were and keep it pointed there.

Once I positioned the flash light, the first one looked like it was drawing close. Then the unbelievable happened - a really bright one came from the right of where we could see the sky and was moving real quick. It passed by the first one we saw and just kept going to the trees to the left were we could not see it anymore. This is when we counted about 5 more little ones coming in our direction - not fast, but steady.

This is when I felt real uncomfortable. They were still reacting to the flashlight but it got too scary, so we went back to the fire. I noticed the first two we saw stayed in the same location we left them at, but I could see a lot more in the skies right above us - but a good distance away. Then the scuriest moment in my life happened. I felt as if we were being watched. I could see light in the trees in the distance and it looked like there were 3 different lights at different spots and they looked like they were scanning the others, but this time they were on the ground. I got really nervous and then I heard the strangest sound ever. It may sound weird, but it sounded like the bugle from the things that were on "war of the worlds". No joke.

At this time, it would have been around 3:30 am and we tried to take our minds off it and just sat around the camp fire. The lights were still there, but were not moving closer. Finally at around 5:30 am and it was getting light out, we could still see some of the UFOs still out there. They made it look like they were just fading out into space, but the ship we saw on the ground with the strange light was still landed and still emitting beams. We could see it in the trees, but it was hard because there were so many to look through.

We decided to walk up to it and get a better look. It had about four lights across in a horizontal line and the ship itself looked invisible except for the lights around it. We tried to get closer and when we did, it looked like it was moving towards us. It shot a neon blue light right above our heads and we took off back to the camp site. It made a strong winding sound like a turbine or something, but I know for a fact this was not of this world.

When it started to get lighter out, we lost track of it in the distance of the trees. This was not the first time I have seen these around here. I would spot them when I would be in town, but when I would show people they would just say they were satellites and I knew they were not. They didn't move as much as when we saw them in the woods, but you would look at them one time and the next they would be in a different spot in the sky and would glow with some type of eternal lights.

There is a lot of military up here and I don't know why they have a lot of choppers and it makes me wonder why. After the night we saw them and were almost like communicating with them, I can see them every night now - just not as close. This was real and I really do think people should have the right to know what is really going on.

Witness #2

I work for a pizza place in St Charles MO. On a delivery I saw two objects in the horizon. When I pulled onto the street near the home, the lights circled around each other and then drifted away from each other. I was in a panic. I stopped at an overpass and noticed there were 4 objects just floating in the sky. I went back to work and told my manager. We both went outside and saw several objects just floating in the sky. Then, we saw MANY unknown aircraft flying around. Most of them were very quiet. They had the shape of F-4's. One of them looked like a 737 but had engines like an F-15. They flew over us at about 1,000 feet or less. They flew around the lights for 20 minutes. The lights just vanished and then the planes just flew away. The planes went east very fast, east of St Charles is Scott AFB.

My manager Brian was a witness and can prove all statements are true. We both think it was UFOs. The planes we saw were not normal aircraft. This can be confirmed by both of us. The lights flashed white, yellow and red. They floated side to side. They were seen to the east of St. Charles, MO near the airport. The total numbers of lights seen was four. When it all started, they moved in ways no human aircraft can move. I *knew at first sight that it was not a normal aircraft. I felt very scared and excited. The whole situation made me really nervous. I knew something was not right. I knew the aircraft I saw were not right. I studied aviation and all this did not make any sense.*

Witness #3

I have made many UFO reports as of the summer of 2013, so my details will be limited. Since first sighting of an orange fireball UFO in July, their frequency of appearance has increased. As of 10/10/13 the sightings have become almost daily. There is a group of four fire ball UFOs that continuously appear in the sky in my area. Their flight patterns and formations vary from day to day but they are WITHOUT doubt some form of aircraft. Their unusual flight patterns lead me to believe they are experimental/extraterrestrial in nature.

Tonight I witnessed what appeared to be a landing in the mountains off in the distance. The craft illuminated the ground below it, but landed below the tree line, so craft details were not visible.

CHAPTER SIX
IN AND OUT OF THE WATER

Witness #1

I was walking on Siesta Beach just after sunset on 1/19/2013 less than halfway between the Terrace and the first lifeguard stand and I noticed an orange glowing object coming up out of the water at about 30 feet from the shore. It ascended into the air and hovered in one spot at a low altitude, then proceeded to move slowly to the right in a straight line or due north. I took a couple of pictures and wondered if anyone else was seeing it like I was.

There was a guy standing a few feet from me and was watching me trying to zoom in for a closer picture of something in the sky, but it seems like no one else was paying attention. When I tried to take the last shot, the object disappeared into the clouds and all I got was what I thought were dark clouds. But when I hit auto fix on my photo editor, you can see traces of the object behind the cloud. I wish I would have been more prepared for this event because I could have shot it on video. But it seemed like I was in a dream state and the event was much longer than the short time span of the shots taken.

Witness #2

I was in the sand and out of the corner of my eye I saw an object that was glowing bright orange red with a glowing tail behind the direction it was moving. It was making a noise that I have never heard. It moved at speeds that were not human. When I saw it I knew exactly what it was. It came from the north of the island coming toward us, bouncing up and down rapidly. The heat from it was intense as it approached, then shot out over the water, descended beneath the water and shot out at a ridiculous speed till it looked like a star then. It glowed brighter and dimmed for about ten minutes. (Seen from Sarasota, Florida Beach).

Witness #3

I and two other family members were walking our dogs down Lakeside Drive in North East, Pennsylvania (a town just northeast of Erie, PA) at around 10:00pm near our home when we saw an orange orb traveling over Lake Erie, less than a quarter-mile from the beach which is located down the lake bank that runs parallel to the street (the bank drops about 30 feet to the beach below).

The orb traveled west to east across the lake and it was flying low at around 200 feet. It was a solid orange glow, similar to a flare but not pulsating and there was no trail or sparks behind it. There was a single blinking white light at the rear of the object, but no visible wings or any other lights that one would see on a plane. It traveled in a slow arc across the lake and we moved to the edge of the lake bank to get a closer look. The sky was growing dark, but was not pitch black in the area we were looking because the sun sets direct over the lake, in the horizon in the direction of Canada.

Thirty seconds later we saw another orange orb of the exact same appearance (orange orb with a blinking white light in the rear) move in the same flight pattern behind the previous object. When it moved across our field of vision, we lost sight of the first object behind some trees along the lake bank. The second object also soon disappeared behind these same trees from our field of vision. Thirty seconds after this second object disappeared, two orange orbs appeared, one following the other in the same direction, travelling west to east. Both objects stopped and hovered over the water. The one in front suddenly lost its orange glow but a solid object remained silhouetted against the sky and changed directions northward in the direction of Canada. During this time, the object in the rear slowly descended to the water below. It landed on the water and stayed there for 10 seconds before the orange glow suddenly darkened and there object sank into the water, leaving ripples behind that were visible by the faint bit of moonlight across the lake.

Two men in a truck pulled up near our road after the objects were out of our sight and told us they were looking for a way to view a lightweight air craft that they were told was being launched over the lake. I believed at the time that they must have heard something on a police scanner about some sort of demonstration over the lake but I have yet to hear anything in the Erie Times or on the news.

After about 10 minutes we saw fire trucks and police cars heading in the direction of Orchard Beach Park, in the direction from where the objects originated in our field of vision. We also saw a Coast Guard vessel and two police boats searching the water where the last orb sank. After using search lights they eventually stopped and focused on one area, the same area where we saw the object sink. So far no one else has reported this sighting to my knowledge.

CHAPTER SEVEN
VISITORS COMING OUT TO VISIT

<u>Witness #1</u>

I have had two very close encounters that I witnessed over my area where I live. I would report these but it's easier to explain.

The first sighting occurred in August just before school started. It was 11:11 at night and two friends and I were sky watching on St. Pete beach. When we didn't see anything, we started to look down and our breaths were taken away as we saw this blue orb coming towards the beach and it stopped just above where the waves were breaking on the shoreline. As we looked underneath it we saw this small almost glass like child sized figure and it paced the beach for about 15 yards till one of us spoke to see if the others were seeing it and it froze where it was and looked at us and the whole time the orb was above it. But as we saw it turn towards our direction the orb strobes were three times so bright that it was unbelievable and after the light, it was just gone ... and we were scared. We sprinted all the way to my car and went straight home and I never remember running that fast in my entire life.

Then the second sighting happened when it was me and my same friend from the first sighting. He, but the way, is always there too. This time, we were sky watching at Jungle Prada and we witnessed these massive orange orbs that came from nowhere. At first there were just two. The third showed up in the blink of an eye. They hovered, rotated and flew over the area we realized was the first place we saw the child figure and blue orb (St. Pete beach). This all happened for roughly 4 minutes till the orbs just dispersed.

Then my next sighting occurred in Seminole and this is the one that terrifies me to the core of my being. I and two of my other friends had just pulled up at a bridge near Osceola High School and we were waiting for another kid to show up. And once again, right around 11 my friend and I noticed that everything was eerily quiet and the street lights started to just turn on and off, but we thought that was normal. But then they started going all spastic, then they just stopped and right as my friend looked down at his phone to say "weird it's 11:11" a very loud rustle and thump came from the water's edge and I locked eyes with a massive being that was 7 to 9 feet tall - skinny, slender and it was dragging a massive bag - one that looked like the kind you put a Christmas tree in. But what terrified me were its pure white glowing florescent eyes and then it dipped under the bridge. Tears still come to my eyes when I talk about this and I'm absolutely serious that I had to draw it to just get it off my mind because it was just burned into me as an image.

Witness #2

My son and three other men were over a mile off the main road on the Cape Fear River fishing all afternoon. Just before dark my son wandered into the woods to maybe spot deer feeding in the afternoon. I decided to walk up the two rutty, muddy roads with thick woods on both sides to the main road because fishing was slow. The guys had a fire right on the river bank and so I told them that I would be back in a while. As I was walking along the road I was hearing lots of scurrying in the woods on both sides I had no flashlight and thought, "man there are a lot of deer around" because this area is known for lots of wildlife. As I made it along, I was unnerved at the sounds in the forest, so I found myself steadily looking to maybe see what I thought was deer - and maybe it was.

The woods were really dark. The road leads to a field which is maybe 200 yards or more wide, just short of the main road. As I came up on the hill to the edge of the field I was shocked to see three large orange balls moving down towards the ground and towards myself and they stopped just over the trees, maybe a mile away or closer and hovered there. I knew it was not natural. I am a commercial pilot and I know aircraft. I stood in shock thinking this cannot be real. I stayed maybe 10 to 15 minutes, then began to feel nervous, so I made my way back to the men on the river in a hurry - mud and all. I then in a panic, took off looking for my son who was 17 years old at the time and as I started into the woods calling him he came towards me totally shaken telling me "Dad you will never believe me, but I have been hiding from Aliens. I am scared they are in the woods." I believed at that time he had seen something - he was serious. I told him "Jr. I just saw three orange UFOs up at the road and It scared me so I came looking for you."

We made it back to the river and my son excitedly began to tell the other three men what he had seen and they started on the both of us. What I did not know was the three of them saw the same thing as my son during that time. The sequence I am not sure of, but we all saw the same. It was a clear cold night. Stars were bright and about that time one of the other men, holding a fishing pole, hollered to look up and to all our amazement, the stars looked as if 6 to 8 of them from different parts of the sky moved very fast - seeming to converge in a group. Then, all of a sudden, three of the objects side by side came over our heads and landed or went out of sight on the other side of the river - maybe two to 400 yards away. All three lights were round, bright, white, blinding lights as big as maybe 100 yards in diameter - all side by side, all touching down or going out of sight in the forest just on the other side of the river.

It scared all of us so bad. The three friends of mine, along with my son and myself dropped our fishing poles screaming "let's go home". We left our poles and were all speeding up the muddy road to the main road - each and every one in a panic. Some of the men were fussing take me home. First, I need to find my wife and children. When we made it to the top of the hill, this is where the trees on both sides end and the field starts, I slammed on the brakes. All of us were in shock to see what looked like a brilliant white egg-shaped sphere with a tail with long spikes all around the middle and front section hovering 20 feet over the main road 200 yards from us and it looked as if it or the spiked part of it was making slow revolutions. This I tell you this is true. Four grown men as old as 50 and my 17 year old son were frozen in my truck looking at this thing straight in front of us. The lights were brilliant white.

The object then appeared to raise up, started down the road in the direction that we intended to go, and then it shot off just over the trees and out of sight. The three orange balls were no longer there. All five of us were terrified - we raced to Donnie's home first. He ran to the door and his wife came out. We all told her of the story and were all looking at the sky. Gene, 50 years old, lived next door. We ran him home and then raced to David's home. His wife came out to meet us and all of us witnessed strange lights over the trees behind his home. My son and I started home to see cars stopped alongside of the road looking at the same lights.

We made it home and were terrified, trying to calm down. We found ourselves sitting in what we call the red room, which is my private study off of my bedroom on the rear of the home, staring out the window at the sky. I live on 6 acres of land and we have a large dog kennel just in the woods behind my home and a Chesapeake Bay Retriever who lives on my back steps - a guard dog. She lets us know if anything from a person to a squirrel is on our property.

We were home maybe 4 hours and calmed down a lot. No TV - didn't have one at that time. I guess it was around 1 to 2:00 am and there was a noise that sounded sort of like a prop jet that flew tree top high right over the house. I ran outside and could see nothing. Within 30 to 45 minutes all of the dogs in the kennel were roaring. They are all hounds and they will tell you when there is something there that is not supposed to be. At that time my retriever was going crazy - hair standing up on her back and barking towards the kennel.

My son was scared but I convinced him to go with me to see what the dogs had stirred up. My son and I opened the door and the retriever took off. You see, most of the time she will not leave the patio unless I go out if she is stirred up. There was frost on the ground, we didn't even have shoes on. We took off behind her and she ran just behind the kennels and all of the dogs were roaring and her hair was standing up and she was barking at something in the bushes along with all the other dogs in the kennel.

Then she took off after whatever it was, and so my son and I ran back down the road that leads to the kennel to my back yard at the very rear of the property trying to cut off whatever it was she was chasing. When I got to the blueberry hedge row, which is at the very rear of the property, she was still making her way to me in a roaring like panic. At that time I stopped, not really scared, but kind of numb to see this maximum of four feet tall creature person/alien staring at me. It looked as if it had a clear glass like covering around it. It appeared to have a faint glow of red and black. On the face I could see nothing but what looked like a red set of goggles and a black (or dark) covering over the lower part of its face - sort of like a mask. The body appeared to sort of glow. I can probably draw it better than explain it. It then disappeared just as my son and the dog got right up on it and my son then told me "you see I am not crazy." This was the same description as he told everyone.

I will not go into detail, but my son and I were committed to an insane asylum by our family because they really thought we were crazy. We were released within two days and then my father came to me and apologized when he said he had heard on the radio the news that lots of people had been seeing strange lights and that the government reported that a Soviet Satellite had fallen that same night.

I am 45 years old with 4 children, a church goer, retired builder and commercial pilot. I have been ridiculed, committed along with my son, and oppressed to say anything, but I know what I saw and four other men will say the same. I have been since hooked on UFO files trying to find others that have had the same experience. I will take a lie detector, be hypnotized – whatever, to bring the truth to the public. My son and the other three men all saw the same and have their own accounts. Thanks.

Witness #3

The events of January 2006 were the culmination of nearly two months of what I believe were some sort of extraterrestrial contact. Multiple UFOs. Entity sighting - very tall approximately 12-15 feet. Due to the extended nature of this event, it is difficult to summarize in one short paragraph. I am writing this because of the recent airing of the Bucks County, Pa. sightings. During my sighting I have also witnessed the same strange twinkling lights descending throughout the trees. This is why I am writing this letter now. I am willing to elaborate more on this and I will also be willing to submit to any testing that would need to be performed to prove my honesty. Many people in my town know of this and I've already been made the fool because of a life changing event that many people don't understand. Thanks "R".

Witness #4

I heard a strange humming sound outside of my house which is on 100 acres of wooded area near Piedmont, Missouri. I went outside to investigate what the low humming was and saw a large brightly glowing orange/red ball of light on the NE corner of my house, just above the roof line and approximately 15 feet from the house. It looked like a solid object underneath a halo of light surrounding it. I was standing about 30 away from the object. I was in shock and stood there, frozen, while calling to my 36-year old son to come out to see it, which he did. We both watched as this object for about 20 minutes or so as it hovered in the same spot, wavering somewhat, then it moved to the Southeast and noticed that there were at least 10 more of these objects in the woods at and below the tree line about 100 yards from our position. The object moved very slowly towards the woods, seeming to be unaware of our presence or not to care that we saw it. I tried desperately to get my cell phone to come on so I could take a picture but it would not work. I thought the battery was drained, but after the object was out of sight the phone resumed working.

Next, we turned to go back into the house and something caught my eye. There was a 4 foot tall humanoid being standing next to the tree and a little behind it. It looked greyish but it was hard to tell in the darkness even though there was a full moon out. I just got a glimpse of it before it vanished. Oddly, I did not react by yelling, or running away. The impact of what I saw did not hit me until the next day. My son said that he did not see it. When we got back inside I was shocked to see that it was at least 50 minutes to an hour later than when we went outside, as it seemed like only 20 minutes of time had lapsed.

Next, we heard a helicopter approach and looked out the window as it flew in the same direction as the UFOs. I have never seen a helicopter out here in my entire life. We did not go outside to look at it. The fact that both of us were able to go to sleep after that was amazing to say the least. It was as if we were in some kind of a trance-state.

The next evening my son was in town getting some groceries when he was approached by two policemen. They asked my son how he was doing and if he believed in aliens or ever saw an alien. He replied that he did not. They again asked him the same questions, which he thought was very odd and again replied that he did not. There had been no prior conversation, and he does not know these men very well so he thought it very odd that they would say something like that. He was afraid to tell them what he saw the night before. This event triggered the memory of the prior night and when he returned home he asked me if I recalled the same things he did, which I did recall, including the sighting of the being by the tree. We both found it extremely odd that we did not think of the previous night's event earlier in the day, or that we were not more upset by it, and had not called anyone about it. How did the police know about this event? We spoke to no one about it.

I called you to make this report in an effort to see if anyone else has reported anything similar, but I don't want any publicity and don't want to talk to anyone else. The whole thing has me a little scared. Note: We have seen balls of light in this area before, and have seen unidentified objects, but nothing as close or as dramatic as this.

My parents and grandparents owned this same land for many years, and they said that all kinds of weird stuff happened here but they did not talk about it much. My mom and dad saw a UFO over Clearwater Lake in the early 1970s. I remember that my *grandfather talked about some missing time and seeing a red fireball back in the 1940s, and that my parents always checked and locked all the doors and windows in the house at night, even during summer when it was cool enough to leave the windows open. They didn't talk about it much but now that I think about it that was pretty understandable.*

Witness #5

We were on a camping trip in the black mountains about 45 miles north of downtown Phoenix. We first felt an abundance of static electricity in the calm night air. We stepped out of our tent to get some air when we saw numerous swirling lights. We were all very startled and stayed close together not knowing what to expect. We watched intensely as we saw the lights merge together and slowly descend into a clearing approximately 75 feet away.

We walked towards the object and saw three beings walk out and slowly look over their craft as if something was wrong with it. They beings looked man-like, but walked in an exaggerated upright way and very stiffly. It was at this moment that we were noticed and we all walked toward the beings against our wishes. It felt like we were on a treadmill being drawn toward the craft. The next 45 minutes are very hurtful and make me sick to even think about. I will gladly speak to someone in person about the experience. We were all violated in different ways.

Witness #6

I'm doing a numbered format that follows your written instructions to write my report so that, hopefully, you can better understand this rather bizarre and unusual event. The beginning of my story is boring, but please keeps reading because the story WILL get WERID, and maybe----just maybe--it can be a significant puzzle piece to the giant puzzle that you all are trying to piece together and figure out. Thank you.

1) I was in the back alley of my house. (I have stray cats that I feed; but they hadn't come yet. I was looking for one of the cats that was pregnant, and I was worried about her babies.) My house isn't far from the back alley of the house. (I'm horrible with feet estimations.) But I'd say the back porch/yard of my house is about—for lack of the proper knowledge—about two in-ground, family-sized swimming pools away from the back alley where I feed my strays. I know it was about 9 pm because it had just gotten dark, and that's what time it begins to get dark that time of year where I live.

2) I noticed the object because it would have been virtually impossible "not" to notice the object.

3) Well, since the object was fairly low-flying, I knew it wasn't a plane or balloon or other such easily-identified flying vehicle. Plus, planes, etc. are not circular/orbs, nor do they glow a fiery, beautiful, bright orange color. So I reckoned it wasn't something that is publicly known regarding military or new private-sector experimental flight vehicles because I just can't see them making a "fireball" vehicle. But what do I know?? SO, I thought "it" was something unknown to (most of) the public.

4) Okay. So, the orb-like bright orange—for lack of a better term—"fire ball" flew away. (By the way, this wasn't the typical "ghost" orb. It was way bigger than those "ghost orbs" folks have taken pics of, yet not super large either.) Anyway, as it flew away, it DROPPED 'something' out of it. I cannot tell if it dropped out/released its object on PURPOSE, "or" the dropped object 'CRASHED', involuntarily into the neighbor's tree in my back alley. The 'dropped' OR 'fallen out' object was either black or dark gray. Now here's the SUPER WEIRD part.... The object that either purposely dropped into the tree or accidentally crashed into the tree was robotic, I assume. I say this because it did something - let me explain. Okay, for describing purposes...picture a black Jack-In-The-Box child's toy for the next story portion...This dark object was not that much bigger than a Jack-In-The-Box, nor was it much different than one!!! As I watched the black object, a smaller—again for lack of better words—strobe-like or snake-neck-like object protruded from the middle of the original small black object that fell/crashed/dropped purposely into the tree. On the top of the "neck" or "snake-like neck" there was an EYE. (Not a real human eye...just something like a robotic eye).

And therein lies my 15 minutes: ME watching the "EYE" watching ME!!! I know it's weird, but it is what it is. Or was what it was. The "eye" scope watched me for about 15 minutes until I realized that my dad would have LOVED to have been witness to this strange event, regardless of what it was (even if it wasn't anything super special like a UFO or what have you - it would have been fun for my poor dad, and it was still a unique experience! But he missed it). SO, upon that realization...I ran into the house (again, a fairly short distance, in that it probably took me about 10 seconds to run into the house)...then another four to five minutes to TRY to explain something so weird to someone so that they understand you enough to really want to come out and SEE!!!

5) HOWEVER, and of COURSE, my poor dad missed it. It was just gone! So I'm SO sorry to disappoint you by NOT being able to tell you how/when it "flew" away!!! I mean obviously it 'went' away from between about four to five minutes----give or take 45 seconds or so.... (I did number 5 and 6 in reverse because it makes it easier for me to relate my "feelings" about it at the END of the story, versus the number five instruction you give. So here it goes...)

6) How I felt. Well, I checked, "affected me physiologically" whilst filling this report because I think I "was" affected, in that I consider myself reasonably intelligent. I would have thought to run and show my dad ANYTHING weird, fun, unique, exciting, etc., etc. Especially a weird, robotic MOVING, small object with a "NECK" and an "observation eye"!!! However, I was mesmerized. I didn't move, sit to rest my feet, move my eyes off it, or anything for the whole 15 minutes. I was like, 'locked-in-gaze' with it (and it me). Like I said, I read a little, but I don't know much at all! Yet, it wouldn't take a PRO to figure out that it was observing me. And, I hate to add this because I'm quite sure it'll make me see 'crazy' or an unreliable witness; but a women kind of knows when she's been stared at or watched. I 'felt' it was definitely watching/observing me, and I was pretty transfixed. It was only when I snapped out of it, that I was able to fetch my dad (too late to see anything). It took me 15 minutes to form my own thought (of my dad).

I don't know if there's any info I might have accidentally forgotten OR anyone would like or any questions. I'm very nice. I like people, and I wouldn't mind answering anything at all. I guess since I'm not allowed to put my personal info, you'd have to contact MUFON and ask them. But I'd sure like to know if anyone's had a similar experience. You hear "UFO' stories over the year; but I've YET to hear one that a person has a close encounter with a ROBOTIC entity! I'd SURE like to talk to someone else if there's someone out there. TOO bad I can't leave my email... Ahh well.... Bright Blessings to you all,

P.S. It was moving its "neck". Like a snake would do to look into a hole or something. And it definitely appeared robotic (not organic entity, but who knows!!!). Email me if you want me to email you my sketch. Sorry I didn't have a camera. I'm not a 'sky-watcher'. Just was looking for my stray cats that I feed. I didn't realize the MUFON form would have an upload option. I wasn't prepared! Sorry! So I'll draw what I saw FIRST THING tomorrow morning (April 2nd), and I'll have it available for you by 5pm EST. if interested. Peace everyone!!!

Witness #7

I had been sleeping when I was awakened by several laser-like red beams which seemed to come from the direction of the TV across the room. The red beams spread out and lit the room with a light so intense it reminded me of pyrotechnics. As the light scattered, it collected in the corners along the walls and ceiling and bounced around (similar to sparklers) before dissipating.

In the next moment I was aware of a tall thin human shaped being to the right of my bed. He was wearing head to toe black with some kind of sparkles all over but not an intense covering of sparkles. As I was processing what I was seeing I was not fearful, primarily due to the sparkles which in my mind ruled out a criminal element. The sparkles could have even been residual from the light that predicated the appearance of the being.

As these thoughts were going through my mind, I hid my face and then decided to take another look and raised my head. He was standing in the same place on the far side of my bed to the right, about 6 feet from me. His arms were crossed in front of his chest but I didn't notice his hands, nor could I see the lower legs. As he stood there with arms crossed, he took a step forward with one foot and turned his head to the left to look at me. I put my head down again and when I looked up a moment later, he was gone. Rather anticlimactic, but I often wonder if he'll be back.

So many questions haunt me with no answers. The next day, as I was trying to make sense of it, there was no question that it was not a figment of my imagination, although I had never heard of a tall, black E.T., at least not yet. So I Googled 'tall thin black alien' and found several similar sightings, but most were in Europe and I am in Texas. One was so similar it was described as having a sheen in lieu of sparkles. But that sighting was 20 years ago and during the day. So all things considered, everything I have learned since my sighting has added credibility to my experience. I only wish there was some way to exchange thoughts with my visitor.

Witness #8

My family lives in an apartment on the bottom of a hill. On the top of the hill is a baseball diamond, and right on the edge of the fence are woods, so we are used to hearing stuff when we go out there. Usually it's just rustling of leaves and animal noises. This past month my mother has been hearing the "DROID" sound usually pretty quiet and far off, like when you turn on a droid phone (5 times in the last 30 days). My mother's name is J; she and her friend K, were out on the porch having a cigarette tonight while I was putting my daughter to bed, so I didn't see this part. They said they were outside smoking when they heard the "DROID" sound, so naturally curious, they went up to the top of the hill to look and see if there was someone walking around in the baseball field.

They said they saw a white softball-sized light, hovering in the air, with little blue and red beams underneath it (kind of like a flashlight with glow sticks hanging off of it I guess) and a few human-like shadows. They are unsure of how many there were, but they estimate 3. They said it looked like there was one on the left side of the light, and one on the right, and about 5 feet in height, but they are unsure because it was far away. They said they saw the things move and sway, but not any extraordinary or fast movement.

Next, they said they noticed a big figure, approximately 7 feet in height and 4 feet, shoulder width, which was standing behind the light. They are unsure whether it just appeared, or if it walked up, or if it was already there and they just didn't notice it. That one did not appear to be doing anything but standing there, watching the other ones dance. J and K said they stood and watched the beings, mystified and silent, for 10 minutes or so, before they realized that they probably shouldn't spy. So, they walked back down the hill, and went inside.

I walked out of my daughter's bedroom, just as they came inside, and that is when they told me what happened. They looked shaken and serious. After a long fight to get the kid to sleep, I needed a cigarette, so the three of us went back out to smoke again. We did not go to the top of the hill to investigate any further. In fact, we heard a noise off in the distance (in the same general direction in which they said they'd seen the things) that sounded like a shovel digging in the dirt. We felt uneasy, like we were being watched, and went back inside. I plan on going out there tomorrow to see if I can find anything that might answer some questions.

Witness #9

I was at my house, and I had been sleeping in my bed. My friend was sleeping over at this time, but I know it wasn't her, because she's really short, and this "alien" was really tall and super slim. No one in the house was this skinny. I had my phone off at the moment, but when I woke up to turn to change my position, it was on, and the light that was coming from it, gave me enough light to see this thing.

It was really tall, about 6 to 7 ft. which no one is around this height in my house. It had a weird looking head, sort of bigger than a human's but not overly sized. Its face looked scrunched up, with its forehead covering its eyes. I saw that it had very long arms that were about 11 inches away from touching the ground. Its fingers were 5 inches away from touching the ground. Its fingers were the size of a banana (approximately 6 inches) and the tips of them, looked like claws. Very, very sharp.

During this sighting, I felt fear. It's like fear took over my whole being. We stared at each other for what seemed to be a couple of minutes, but when I looked at the clock, it had only been a few seconds. A couple of weeks, after this incident, I got severe depression, blocking out everything. I think I scared it off, because when I woke up I jumped a little and stared at it, then it went away. In the morning of this incident one of my hair extensions was gone. I think it thought that it was my hair to do some testing on.

I don't know what it wanted to do with me. I feel like it wanted to abduct me, but I woke up, and it didn't want to take me when I was awake so they took my hair extensions instead. I would really like an explanation as to why it was in my room, and what it wanted with me. I still feel like this sighting of the alien is the cause of my depression.

CHAPTER EIGHT
AROUND THE WORLD

Australia

It was a beautiful and sunny day. I was playing at the front of the house, with a friend's child. There were approximately 5 other people in the street who observed this event. The child alerted me to look up. From the south I noticed a large orange orb. It looked like a small sun, glowing. It was bright orange, but did not hurt your eyes to look at it. As it neared and became close and above me, I could have hit it with a rock. That is how close it was. From out of nowhere, a smaller orange orb appeared. Side by side they moved off, towards and over the ocean. They flew past the naval base, on Garden Island. They continued nonstop, until I could no longer see them on the horizon. I have seen these orbs again in Sydney Australia, years later. Four of them on this occasion, they flew in a straight line. Then they changed course for space.

France

Hereunder is the sighting and photos of a French family who were witnesses to a UFO sighting near Geneva from the French border. Translation is my own.

I have joined yesterday the MUFON network and do not know precisely your modus operandi yet. A minor took these photos and she and her family wish to remain anonymous. Photos when zoomed up show six orbs forming a second light which were not visible to the eye when the sighting occurred.

Bonsoir John, Good evening – John, here is the story of our sighting of Sunday Dec 7th 2008 around 5:15 PM (photos show 6 15 PM as hadn't changed to winter time yet). As I live in the high Jura Mountains, cell phones cannot go through well. So I went to my large window to receive better - my window is facing full east. That is when I saw at a distance a big brilliant light above Switzerland (difficult to evaluate distance). After hanging up, I called my family so they could watch too and went to get my camera (APN with 70 300 mm lenses 6megapixels).

For us this light couldn't be a star (too big) and too luminous. We thought it could be a plane (as Geneva s airport is close by), but the light wasn't moving and was too brilliant.

We are used to seeing plane lights including the very powerful beams and they are not that big. Besides, the sky was crystal clear, no clouds and under freezing point temperature, cold and dry.

I went out on my balcony to shoot the photos but missed the first one (not a professional) so I changed area, put on my APN (digital camera) and shot two more photos automatically. We then looked a while, then the light disappeared in a fuzzy unequal orangey color, came back as we saw it again, then it disappeared fully.

When I speak of the light, it is the larger one. It's when photos were loaded on my computer that I noticed there was a second one (bizarrely shaped as when fully zoomed one can see a few smaller lights). There were four of us that watched this event and none of us saw the second light. Both photos were taken at 14 seconds apart ... for us the light didn't move (we hadn't seen any movement) and it's by seeing the photos on the ids forum with an axis traced that we saw movement. I looked tonight at the same hour and saw nothing. The orientation of the photo is full east with north to the right hand side and left being south. The wires seen are electrical.

Photos were taken with the 300 mm night vision. First trees are about 500 meters from my house. At your disposal for further details. I live in Prenovel de Bise Jura, France. Salutations.

Ireland

I was coming out of my local shop when I saw an orange glow. At first I thought it was a lantern, but at second glance it seemed to be moving must faster than any I had seen before so I kept my eye on it. At this point it was heading north in a straight line. It was about 1 mile from where I was. It was just dipping in and out of low cloud cover. There had been some light rain half an hour before - normal weather for Derry City, Northern Ireland.

As I walked back to my house it seemed to stop and hover for about 10 seconds, then headed of at the same speed as before only in a north east direction for about 30 seconds and then stopped again suddenly, then vanished into cloud cover. This was not a ball of lightning, a plane, helicopter, or lantern. There was no sound. It was not a star or meteor or anything like that due to change of direction.

England

My daughter and I were in the garden when a large glowing sphere approximately 3 m in diameter came over our neighbor's wall and descended a few more feet into our garden. It was less than 10 ft. from the ground, moved very slowly and stopped suddenly, and just waited there for a couple of seconds. Then it shot straight up and changed angle of direction. The object was silent but was clearly powered and able to reach speed faster than a plane. It was solid looking.

My daughter was within a couple of feet of it and it gave off no heat or sound. She was closer to it than me. She thought it was metal. There was no visible opening or window - it was just solid and glowing. We both wondered if it had stopped to watch us. That's what I felt - like we maybe were being observed. I know it sounds crazy, but this was not a lantern or weather balloon.

India

Flight "Spicejet" No. 013 from Bombay to Dubai - Departure 19:45 +5.5 GMT/ 18:15 +4 GMT - Object sighted at approximately 20:45 +5.5 GMT - Duration of sighting approximately 1 Hour 30 Minutes.

Object encountered over Arabian Sea - Object disappeared from view when aircraft turned left at coast of Fujairah - Sighting confirmed by member of cabin crew - First such sighting in 45 years of air travel - Object appeared to be a glowing fireball and quite obviously larger than stars - Object distance appeared much closer than stars - Object appeared to have erratic movements: Moved in a straight line, stopped abruptly, hovered and changed path with ease and speed.

Circled aircraft - Maintained speed of aircraft - Easily overtook aircraft and came back alongside - Object randomly flashed light rapidly - No adverse physical or psychological effects felt during or after sighting.

Canada

On December 25, 2013 at 19:15, I was at friend's house when he screamed to get outside now. While letting his dog out, he saw glowing orbs in the sky. We observed 5 orbs in a 'V' formation moving slowly from north to southwest low on the horizon with no sound. They appeared in our view over a high privacy fence and after the first 5 passed at about a mile away, more appeared following the formation. They were single file at the same altitude, not evenly spaced, some closer to each other and some further apart.

I ran into the house to get my phone to take pictures. All that appeared on my screen was a tiny white spot which I will download and have someone more experienced look at. We both knew that we were viewing something unnatural and were yelling and hoping neighbors would come outside as well. There was no sound and no exhaust or blinking lights like a plane has. We are situated 20 miles from Pearson International Airport and these orbs were not going in that direction. We both stood in the snow without shoes to watch until the last orb appeared, then disappeared from our view into the low level clouds and this observation lasted about 4 minutes from the first formation to the last orb moving out of view.

It was the most amazing experience, unlike anything else and I want to see them again. My friend worked at the GTAA and immediately checked Flight Tracker 24 on the internet and radar only showed two commercial aircraft in completely different locations at that exact time. My friend has always laughed at the thought of UFOs, but after witnessing those orbs, he has completed changed his outlook. On my mother's grave, this event happened as I have described.

China

I was playing Monopoly with my little sister inside the house and my dad was outside exercising, when he came into the house looking excited. He told me he'd seen UFOs. He showed them to me outside the window in my room. When I saw them, I knew they were UFOs - they looked very clear. I ran out of the house with my little sister and my dad. The objects were still there, hovering in the sky. They were in a triangular formation, and the smallest one that looked like a tiny sphere slowly came kept moving closer and then farther from the largest one that looked like a triangle. The triangular UFO flashed all sorts of lights very quickly. The small one also gave off different colors, but it changed colors slowly. The one on the bottom of the formation was shaped like a disk, and it gave off purple and blue lights. They were close to us at first, nearly overhead, but they slowly flew Northeast, still in the same triangular formation, until they were out of sight.

We tried to take pictures of course, my dad using his digital camera and me using my phone. We couldn't get any good pictures except for one, though. They were too far away. For some reason, many abilities on the camera on my phone couldn't be used, such as flash and color effect. They worked fine when the UFOs went away.

At first, I was extremely excited. But later, couldn't stop staring at the bright lights and I was so distracted I forgot didn't remember to check the time until 8:20. Luckily, my dad remembered the time he first saw them. A few people passed us, and they looked up to see what we were trying so hard to get a picture of. Some of them gave us funny looks, as if trying to take pictures of a UFO was the most unusual thing to do. Nobody else got excited about them. I wondered if any of them were aliens in disguise.

Argentina

Today 05/04/2013, I saw from my house and witnessed with my wife and kids a flying triangular object with 12 lights around the edges which was moving through the night sky.

After a while the shape of the edges moved, and looked like it was 12 light balls moving in the same direction. Then appeared another one, and after 10 min appeared another one with a really disorganized shape. After 1 minute or less, the light - the last one, came to a line and moved to the downtown direction.

I saw also too many choppers around the area. We leave near a military base. And I think they saw the lights as well. I am sure this wasn't any type of globe object done by humans. They fly very fast and move their direction very smoothly. Hope somebody else sees this in my area.

Mexico

While watering the lawn at night, three objects in the sky caught my attention. As soon as I saw them I knew this was something out of the ordinary, and immediately turned off the lights to get a better view. The sky was very clear. Three red-orange orbs ascending in the sky to the south southeast in a rather straight path. When they reached a certain height, each one faded out and began moving in an arch pattern until you couldn't tell the difference between the orb and a star.

A few minutes later, another three objects ascended one by one, again fading out after and mixing with the stars. After witnessing the first six orbs I went inside to call my dad. When we returned, we witnessed the same pattern repeat with about another six to nine orbs with the same characteristics. After a little over 20 minutes it stopped and we were left scratching our heads.

Norway

Kleive, Norway, Scandinavia Sunday, 25 January 2009, 0:40 a.m. My father saw an orange object flying in an east to west direction. He then alerted me. The object flew very slowly, straight above Kleive (small village). No sound could be heard from the object, when I went outside to get a better view. The object flew in a steady horizontal path and at an altitude of circa 500 meters.

Suddenly the orange light dimmed and the light on the object was turned off or it disappeared into a cloud. No blinking was observed on the object. I noticed that the neighbor's dog barked intensely, as long as the object was in sight. The dog stopped to bark, when the object disappeared. The duration of the sighting was approximately 3 minutes.

CHAPTER NINE
PRE-MUFON STORIES

1968

My husband and I have wondered for years if someone else may have witnessed this same huge orange pulsating sphere. It was the Friday before Easter, 1968. We were traveling to NM to visit relatives. We left Oyden after work on Friday. We traveled in the evening because traveling with 2 little babies was easier at night. We were traveling south and were approx. 20-30 miles past Moab heading to Monticello. We were the only car on that stretch of highway in the Church Rock / Looking Glass Rock area of the desert. It was dark with no other lights in sight except our headlights.

I'm estimating it was about 11:00 PM when all of a sudden the whole surrounding area of the desert lit up like daylight. I don't know where it came from but we could see through our windshield a huge orange pulsating light hovering right above us. It was about 30 ft. in diameter and approximately 20-40 feet above us. I remember telling my husband, "stop, it's going to hit us". He braked and we just stared at it for what seemed like an eternity, but it was probably only a minute or two. We did not hear anything or see it coming at us, all of a sudden this huge ball just appeared above our car. After a minute or so the light/sphere just sort of bounced over a couple of times to the East and disappeared over a mountain.

The only comparison of what we saw that night would be like a bright orange sunset in the dimensions I described. I remember we were very scared and drove to Monticello, the nearest town and parked in front of a downtown business and waited for dawn to continue our trip. We were just a young married couple with two little babies and it never occurred to us to stop and report what we saw. We have always wondered if anyone else saw this.

After this, my husband and I became very interested in UFOs and to this day I wish I had documented every detail and the exact time everything happened. My husband disagrees with the time of sighting and thinks we lost some time that night. A few years back I wrote down everything my husband and I remembered from that night. This is the best recollection of this event to my knowledge

1967

One summer night, I believe it was 1967, my girlfriend, Debbie H., and I were sitting on her garage roof, talking. We were looking over the fields behind a row of houses on N. Leroy St, MI. All of a sudden, a bright sphere came up over the tree line. The tree line was less than 500 ft. away. The Sphere moved very quickly, with no noise and came to a hover, two fields away, about 250 ft. It hovered over a grove of pine trees that Johnny (of Johnny's Nursery) had planted for Christmas Trees. Then it descended straight down and appeared to land.

The pine tree grove was lit up from where the sphere landed. We got off the roof and ran down the street to the home of a Fenton policeman. He told us to go home and stay away from the grove. We did not go near that grove for at least a year. When I did go to inspect the landing area, it was devoid of all growth, just gravel. Over the years, I checked that area, and nothing ever grew there. Additional information requested. We did not know what it was that we saw. We were scared.

1966

I have entered reports before. Times and dates questionable or may vary. Only EAFB knows for sure. Good luck trying to concur any information. Four on our watch. EAFB teletypes ring off with bogey accounts of cluster-like object that deployed or changed shape or?? Received messages for hours and hours of stat reports. Others joined us on our watch shortly (about) one hr.(never happens) into reports and relieved us of our duty with AFB teletype monitor, so we were freed up to handle routine traffic from various White Alice sources, Hono, Japan, and lower 48. Were advised of our mission as personnel pertaining to all details of security clearances and some, zip-lip crap from some lieutenant with bad breath, so we never discussed any of it.

I for one never even wrote home about it and frankly zobbed the whole idea. I liked to bivouac on my 72's and witnessed a light, bluish hazy oval shaped object over Buskin Lake some months later at dusk. Watched it, no noise etc. and just laid down in a wrapped poncho and went to sleep. Didn't report it. Who to? Military authorities? I just can't understand how "authority" goes day to day with whatever knowledge "they" have about this stuff in an aloof, unattached fashion until the rubber meets the road.

At this point, any attempts to mention this to family or anyone at any point is met with harsh contempt because I waited so long to say anything. Who am I? Clark Kent? Thanks for the memories.

Want to hear about missing time? X-file stuff? About five or six years ago, I started getting little mental snippets of being somewhere not my home (in 1953?) in Portland, Oregon in a dark place sitting closely to an extremely beautiful blonde lady who was more comforting than anyone I have experienced. Across from us in the dark haze was a military guy in khakis sitting silently. These are the earliest memories I have in my life, oh yeah, something went up my nose, I could "hear" twisting and grinding and pressure but no pain. I am sure that the blonde lady was there to comfort me. There was no panic or trauma and certainly no memory of it until the above mentioned time. Why the hell would I remember things like that? You're in the business; be my guest.

<u>1965</u>

I was on watch at the USCG Station watchtower at Yaquina Lifeboat Station, Newport, OR. The watchtower is located at the present Lighthouse Museum and parking lot North side of the Newport State Park, Newport, OR. The sightings began about 2300 PM. I was not the person who saw the objects at first, we are mainly concerned during our watch time for things at seaward (west) and at the inlet of Yaquina bay (west) in case of any emergency that may occur that would require the Coast Guard - needed in the event of rescue efforts.

There were people parked at the parking lot below the watchtower who started seeing the objects in the eastern horizon. They banged on the tower legs and got my attention. They were asking me if I was seeing them. I said no, but I would scan the eastern horizon to try and see what they were referring to. It took a few moments before I spotted the objects and then I called the XO/CO at the base facilities to report what I was observing.

The CO came to the tower and observed what was going on, left the tower and said that I was to stay on watch until notified to do something else. My watch was to be relieved at 2400 hours, but I elected to remain with the new watchman to observe the objects that were in the eastern horizon.

Apparently the CO called the Air force to investigate, jets went to the area and the objects disappeared when they were in the area but returned as soon as the jets left. Some were of yellow color others were an orange color. Some were larger than others and they move back and forth in the sky, staying in the eastern portion of the Newport, OR area from the Park area I was at.

When the jets went in the general area, they went out one at a time, returning nearly in the same area after the jets went away. By early morning, around 0200 am, there were quite a few people in the area, parking lot and other viewpoints to see these objects. There were as many as twelve and as little as three at a time in the eastern horizon, changing directions at various time for whatever reason. The news media in Portland, OR broadcasted the event that late evening which several people drove to the coast to observe. The incident (and the increase of people I saw after 0200 am in the general area) lasted until nearly 0500 am, when one by one they disappeared from view.

It was amazing to me how long they stayed in the area. I added everything I observed, as did the other watch personnel in our log books. Several days later, Air force personnel asked us to write down all that was viewed on prepared forms. A few years ago I went back to the USCGLBS at Yaquina to see the entries, since I am a retired USCG and asked permission to see the old logbooks, which was granted. They were missing from the records they still had at the station. In fact, a period of several months before of *all log books were gone and no explanation of where they were.*

I was able to get news articles and a few people remembered the event...it was a night that has never left my memory and I still cannot say exactly what it was. Since I observed many things during my career until retiring, it was only one of two events that I cannot explain away as normal objects I encountered during my lifetime. They were not normal in any sense of the concept of things, but an event that will never leave my being.

1960

I was at Low Field Ft Rucker, AL. Working outside at night as an aircraft mechanic. 2- Had no reason to look I just did and saw it. 3- Had no idea then and still don't. 4- Amazed and left with a great amount of wonder. 5- It came from an east direction going in a west direction. Moving so fast it was difficult or next to impossible to show it to someone else. It moved in a straight line, at a very, very, high rate of speed to say the least. It appeared to be about the size of a soft ball maybe up to a volley ball, Red-orange in color, with a long glowing red or orange tail. The tail looked like red hot sand cooling off. This lasted longer than the time it took the ball to cross the sky. This occurred 3 nights - one after the other, but from slightly different directions.

1960

The following event happened when we lived on Kaer Avenue, Red Bluff, CA and I was just leaving grade school for high school. This occurred about 1960. I was around 15 years old (I am now 66) and my mother and I were in the house. Mom was probably fixing supper and I was more than likely watching TV. My father came into the house and told my mother and me to come outside and look at something unusual in the western sky. We followed Dad out the front door and into the street on the south side of the house.

We observed a basketball size object (reddish orange) hovering in the western sky. (Note: It was not the moon, as the moon was in the eastern sky just where it should be, with its man in the moon shape, etc.) I had no idea what is was, and my father who was a WWII veteran was completely stumped. This is where it gets weird. The object did not move up or down, left or right, but hovered in exactly the same spot. Then it just disappeared about 7:30 pm. We talked about it the rest of the night.

The next day my Dad came in and said it was back. Sure enough, at just after 7:00 pm it was exactly where it had been the evening before. Same exact place. My Dad called several neighbors who came over and I remember them getting into all kinds of theories as to what is was. Then they observed that if you walked down the street, (heading south) about 50 yards, it disappeared and when you backed up a couple steps is came back into view. The same thing happened if you walked up the street (heading north). We found this out because some of our neighbors could not see it as they walked toward our house, and then it appeared suddenly and if they backed up it would disappear. There was only about a 100 to 150 foot area (window) that it could be seen.

This event took place for 5 nights, and then stopped. The time for each event turned out to be from 7:00 pm to 7:30 pm.

I do remember that on the last night, our driveway was full of law enforcement vehicles and even a couple of National Guard rigs, and a ton of people from around the neighborhood and from town (Red Bluff, CA). Our neighbor across the street, the Red Bluff High School Science Teacher, was completely stumped. On the third night he called some of his colleagues from a university in California and they were there on the fourth and fifth nights of the sighting. Some had gadgets to measure light, and heat, and size. Other gadgets, I do not know their purpose. I do recall them being very confused as to why the object would disappear when walking north and south and even when walking away from and toward it. This was truly an unusual event.

Extra notes: SIZE: The size and shape was the same as a basketball that was only about 10 feet from you. Very large and if looking at a basketball only 10 feet away it appeared to be about 8 to 9 feet off the ground. That will give you an idea of the angle from my view point as I was only 5'6 back then. DISTANCE: Unknown. It did not get any larger even when looking through binoculars. It stayed the same size, and no other features were visible either. The color always stayed the same. DURATION: Only 30 minutes of viewing each night. From 7 pm to 7:30 pm each evening for 5 nights.

Was it there the night before my father noticed it the first time? We do not know, it could have been. It never came back again, and I spent days, months, and even years looking west from our house for its return. I know that it's been many years since the event, but I have to know if others have seen a similar thing hovering in the evening sky. Remember, it was not the moon, a star, street light, car headlights, house light, flashlight, lantern, jack-o-lantern, etc. They could not tell us what is was then, maybe you can now. I am sorry there are no photos. I recall (or think I recall) someone saying that none of the photos would come out or develop. There was just nothing there, only the western evening sky would show.

PS: It was your television show that made me write this. OK, MUFON, tell me what it was, or was not as the case may be. DO YOU CONTACT US (OR ME) OR DO I STAY IN LIMBO FOR ANOTHER 66 YEARS. Thank you James. Please call me ... This is driving me nuts.

<u>1958</u>

On a summer's night in July of 1958 my brother and I were outside playing in our yard. I was 12 years old and he was 11 at that time. It was a clear evening and the sun had set below the horizon. Our house was located on the east side of a paved road. To the west of us across the highway was located a plowed field about 300 feet wide and a quarter of a mile long. The length of the field ran parallel with the highway. On the other side of the field the terrain was raised and was a pasture for the farmer that lived on that side of the road. A very few trees grew in that pasture so we could see to the west of us and had a good view facing west across the highway viewing the sky at about a

15 degree angle with little obstruction. Since the sun had set below the tree line and the sky was not yet black, we had a very good view of the event.

We observed a spherical to cylindrical orange to red object that seemed to float overhead that was approximately 1,000 feet in the air and appeared to come from the northeast at a speed of approximately 20-30 mph. The object was totally silent. The object appeared almost straight overhead of our position and moved to the southwest over the neighbor's field and pasture before finally disappearing over the tree line. The sky was perfectly clear and there was no wind. We had an unobstructed view of this object for approximately 30+ seconds.

We went into our house and reported what we had seen to our parents. I remember this event like it was yesterday. I can see the object in my mind as it floats across the sky and I can see my brother and I standing there watching the object. I remember the conditions of the day and the absolute quiet of the evening. I went to a family gathering this past summer and mentioned this event and my brother mentioned that he too remembered this event and he described the event as I am reporting it.

I am level headed and am skeptical of many things. I served in Vietnam as military policeman and am used to hostile situations. This is my story. My brother is a witness.

1952

First of two sightings,1952, riding a bicycle in rural area 1:00 am 12 years old heading home across a field, sensed something, looked up and saw a fireball, slow, orange, red, with blue/white edges. Softball size, south to north. (Near lat. 33.)

My cousin and I were scared. We went home - mom wasn't interested and we forgot about it.

Second sighting, 1962 approximately 1-2:00 am 1/4 mile from first sighting. Leaving mom's house, heading for El Cajon, CA. about 15miles away. On gravel road on a high ridge going east, girlfriend says to look - a car's lights coming up the valley, below us. Not many cars out there at that time. Watching the "car" lights a mile and half away down the alley, we saw that the lights weren't on the road but near it, rough brush canyon lands. We were going east along a ridge, light heading east down below in the valley. We continued east then down off the ridge to the north. Turned onto a paved road going west towards the light now at our level. Less than 1/4 mile away over a canyon on our level, hovering.

I pulled over and stopped, the two girls were screaming - wanting to get out of there. I rolled the window down, no sound, turned the lights off, no reflection. Girls screaming, I said, "I'll flash the lights and call them over". I flashed the lights and the bright white light, like a street light but way brighter, started moving back to the west on the same path.

We followed the light at about 45-50 mph, we were on a curvy road. It was moving in a direct line, we went around a hill and the light was about 300 ft. high now. We were going down into the valley, the light stayed level. Another 1/2 mile we were 1 mile behind the light and light was about 6-700 ft. high. Another hill to go around and we were farther behind. We lost sight of the light for about 8 minutes until we reached a high ridge about 8 miles from El Cajon, the light was over El Cajon. We were able to watch the light most of the way back to El Cajon

1946

Round colored lights, merging and separating seen in the sky in a parking lot. I was 5 or 6. I pointed them out to my parents, they did not look. They said it was the power lines. I could see they were not near the wires.

CHAPTER TEN
DOES THE GOVERNMENT KNOW ABOUT THESE ORBS?

Witness #1

I noticed the first craft at 2155. The object appeared at first to be a fireball moving above the trees. Once it came into view, it was very clear that it was an orange orb with a spherical structure in the orb.

The object appeared to fly just above 500 feet and at a rate of 30-50 miles per hour. The object was flying in a linear path for approximately two miles and then turned to the east and started a rapid ascent at 60 degrees until it blinked out. This was followed by a second craft with the same description and flight path. Again once it turned east, it did a rapid ascent. A third object then appeared. Once again, it was following the exact flight path of the first two objects. This object, however, dropped a number of flaming objects from beneath the craft that blinked out before they dropped out of sight.

Over the course of 46 minutes, 7 objects followed the exact same flight path. Once the last object passed, three military planes came into view. Two descended from approximately 20,000 feet and came from the northeast and then banked east. The third plane also came out of the northeast, but was under 5,000 feet before becoming obscured by clouds. There were a large number of witnesses to the event, though I cannot give an exact number. I did get a very short video of the last object with my iPhone and it does show a jet coming into view, but not much is visible because I was trying to drive.

Witness #2

I was sitting on my balcony having a cigarette when I heard a helicopter and looked up. It was coming from the north flying low, then slowly flew upwards to this flashing object. It then was nose to nose with the object and stayed there for over 10 minutes. I expected maybe it was a plane that had been flagged or something, but then the helicopter flew to the south where I figured it landed at LAX and the object went off to the east toward the San Bernardino Mountains. I'm telling you it was an extremely strange occurrence. Oh and by the way, while this was happening, all of a sudden all these sirens started to go off, police and ambulance, so of course my mind went to - they've gone emergency mode. I called the airport police, so I'll probably get a visit, but I wanted you to hear my story. Wowwwww!!!

Witness #3

We watched many orbs turn into many formations. The military has cleared my airspace for the day and night and I know why!!!! Stop chasing these things with fighter jets continuously over my neighborhood!!!!!!!!!!!! This is a residential area and we are tired of jets screaming past our house after these crafts!!!! Go practice over the ocean it is 2 miles away, it is unsafe, unpractical and unethical do being such matters over a major tourist destination.

I have been recording all this for 6 months, I know the air force has no reason to be holding training exercises continuously over my house. I wouldn't be upset if it were just a single engine plane, but this is not and we demand a valid logical explanation as to why our military would be putting my family in danger. My wife and kids are afraid to leave the house.

Get your act together, MUFON included, I have been contacting you and the state director since July, and all I've heard is - spent flare casings, do they fly into and from aircraft, or commercial jets, etc. I have video for all!!!!!

If you are testing, go to a military test site or the ocean or your base, and don't drop waste on my housing area. I have hours of footage of that day and night. Do people want to see these orbs going to and touching commercial jets? Please just stop or call me and guarantee my safety, no way you can!!!!!!!!!!!!!!!!!!!!!!!!!!!!!

Witness #4

I was sitting in the living room and my mother was in the kitchen and she yelled out that she heard a helicopter outside. I didn't think anything of it because I thought it might be a life flight helicopter, but when the windows and the pictures started to shake on the walls I jumped up and went outside.

When I looked up I saw a very large helicopter, military grade, flying low and chasing what looked to be an orange light. I yelled for my mother to come out and when I looked back the light and the helicopter were changing directions. When the helicopter banked, the light went from being in front of it to being under it. The orange light then took off fast and the helicopter was right behind it.

We came in and called the hospital to make sure it wasn't life flight and they said it shook the windows in the hospital and that the ER called to ask if the front desk knew if there was a life flight being called in. The nurse told them no. the ER then said they thought it was going to land on the roof of the hospital. The nurse even stated that it did not land at the airport across the street. At first I thought it was a plane being chased but there was only one light and it was orange not the red or green that are fixed to the wing tips.

Witness #5

My friend and I were watching a movie, I can't really recall what movie, because I have been hesitant about posting on here for the past two years and I'm not at liberty to say why. However, I can tell you this is not the first time unusual things have occurred around me such as this ... Anyway, we were watching a movie late at night, and I had to pause it because we couldn't understand the dialogue due to the volume of whatever was outside.

As I said, we thought a plane was about to crash outside so we ran out to see what was up. We made it to the front porch and stood there for a second, the sound had dissipated completely, slow at first but just disappeared by the time we made it outside.

We saw a helicopter about a mile or two away with a searchlight on looking around towards the ground, and I remember thinking it was probably just someone the cops were looking for. Shortly after, I remember everything vibrating like I imagine an earthquake would be like - I mean the ground, puddles of water, windows and cars, etc. We both kind of stopped and looked up simultaneously due to a sharp gust of wind that swept the area around us. Mind you, I'm on my front porch! I look up and I see a huge light with four lights around it, and I can't even begin to account for how many lights surrounded the exterior. I'm not epileptic, but my god - I felt like I could have a seizure staring into this, whatever it was I was staring at. I could make out a definite shape that was a circle or oval. It was all black, the lights illuminated so brightly it was difficult to make out much else of the craft, but the sound I'm telling you was deafening.

I talked to my neighbors and NONE of them heard anything throughout the night. Only my friend and I heard it. How that's possible in itself, I can't fathom; let alone, the fact that this thing, which I'll remind you yet again, was right above my house! And by the way, its radius far surpassed that of my house!

I'm no scholar or anything but this thing's circumference was at least four-seven times the size of the perimeter of my property. So if my estimations are correct, this thing was a couple hundred feet in diameter. And it lingered above my house – no, it lingered above my friend and I - for at least a solid sixty seconds before slowly hovering down my street, until it was above the baseball field at the end of my street and just turned to the right from where I was standing. Then it moved a little bit more in that direction and then just flew off and it was gone. I mean I had my eyes zoned directly on this thing the whole time, and it just disappeared.

My friend and I didn't even talk about it until about a month after it happened. That night, afterwards, we just looked at each other and she said simply, "OK, well I'll see you later" and I saw the look on her face, and I told her to get home safe and call me when she got home.

I stayed outside until she called me, it took about thirty minutes, but within that time I saw and heard at least four more helicopters. Then she called me to let me know she was safe and we still didn't talk about it. The week after that, I saw several military choppers surveying the area frequently and I mean almost every day for a good two-three weeks, not to mention small regimens of soldiers here and there. A lot of similar stuff has happened in this area in the past months after this incident as well.

Witness #6

Driving on route 8 towards B-pt. when my wife and I noticed an object in the sky. It looked like an orange teardrop and all of a sudden a bright white orb appeared from underneath it. When military aircraft (F-16's) approached it, (I estimate about 12 aircraft) it disappeared. When the jets were in the distance it appeared again and we noticed the jets turning back towards the object once again. The sky was filled with contrails!

I would like to know if, or how many, people sighted this. I contacted B-pt. police and they said one other person had reported unusual activity in the sky. We had to exit the highway and could not observe it anymore after that. If anyone else saw this would you please contact me. This isn't the first time for me.

Witness #7

Nine people at a beach house near the Carolina Yacht Club in Wrightsville Beach, NC saw a ball of reddish orange light blinking erratically, silently in the distance over the ocean. A small white light appeared out of nowhere, close to the object, then disappeared. Then about 2 minutes later, it reappeared and we saw it divide instantaneously into nine orbs, assembled into a horizontal V formation, after which it disappeared in a couple of seconds. There was no sound and no other aircraft around, and it was about 10 degrees over the ocean's horizon, about 1 to 2 miles away.

We were all shocked and we all agreed that what we saw was not man-made. The movement was too fluid, and the speed at which it divided was extremely fast, and yet controlled. The sighting was at 9:35pm, and then at 9:37pm.

Shortly afterwards, we saw helicopters descending on the area over the ocean where we witnessed it. At around 11:45 pm, two fighter jets flew by, parallel to the coast. Why were they flying so close to a residential beach town at that time of night? The next day, military helicopters flew up and down the beach all day long, nonstop.

Witness #8

I took a gal to the movies and into the show when management said we were being flown over by a UFO and it was not a hoax. There was a large military and police presence. We then were told to leave in a single file. As we left, the military told everyone not to speak about this to anyone. The object was large and disk shaped. It moved slowly over the place and it turned around very slowly as it went from east to west. Then it went over the airport tower and hovered, then it turned into a white ball of light and a small red light came out of it and flew around very fast. The object then flew straight up and out of sight.

I told the gal I was with to look, but she was so afraid that she put her head into her lap and cried. I said this was a chance of a lifetime and we might never get another one. It was like two plates on top of each other with a top that looked like a Reese's Peanut Butter Cup turned upside down. It had windows which were lit up and there were beings looking out. The craft was about 50-60 feet in the air.

CHAPTER ELEVEN
INTERACTING WITH THE ORBS

<u>Witness #1</u>

Spotted orange light hovering in sky, went into house grabbed laser, went back out and shined laser at it since there were no blinking aircraft lights on it. It disappeared.

<u>Witness #2</u>

My wife, a friend, and I were surf fishing at Topsail Island. I had reeled my line in and just happened to look down the beach to the south when I spotted an orange sphere traveling north in a straight line down the path of the island. My wife turned and saw the object. When the object got directly overhead of us, it stopped. I had a flashlight and blinked it off and on and the object did the same thing. It sat there about thirty seconds, got brighter and just blinked out.

We were still looking for the object and looked south. Again, there came another orange sphere. It slowed down in about the same area and stopped again near where the first one did. I blinked the flashlight again and it blinked also. Started going again toward the north and just blinked out about the same area as first.

CHAPTER TWELVE
BEAMS DOING THEIR THING

Witness #1

I was out in the barn, feeding my pigs. I saw a "comet" and it started sparkling in different colors, then I knew it wasn't a comet. It looked like a disco ball with all of these light beams coming out of it. I don't know what those things were used for, maybe to stun its prey (my cow Queen).

It had this blue laser beam that sucked up my cow. They only took one, but they didn't bring it back. I don't know what they did to it, and I don't think I want to know. The whole thing scares me. I thought for a brief second, maybe they are nice, but I didn't even want to try. I don't want to end up like Queen you know.

Witness #2

I went outside to smoke a cigarette around 11:30 at night. I then saw a black object hovering in the sky and I rushed to grab my cell phone to try to record it. For some strange reason my cell phone wouldn't work. I looked up and noticed a beam of light shooting out with orbs straight to the ground with theses tall skinny beings. I wanted to go inside but something was prohibiting me from doing so.

Now the orbs were what I would think to be scanning the area while the beings seemed to be instructing them with some black device. Then they made their way towards me and I was beside myself. Now there was one that was slightly taller and seemed to be in charge and somehow communicated with me telepathically, saying that they mean no harm and that they have been studying man for centuries.

Now one of the orbs scanned what I can explain as a scanner - like you would see in a science fiction movie. They were examining me for about 5 to 10 minutes and wanted to go inside but I pleaded with the one I was talking to not to go in to protect my family. The strangest thing happened! He seemed to do a slight bow of respect and slowly walked off my porch and left!

CHAPTER THIRTEEN
ORBS – UP CLOSE AND PERSONAL

Witness #1

I had just gotten out of work. My routine was to get 1 can of beer and go to Pilcher Park Forest Preserve to meditate. Not sure of the exact date now, but it was around the middle of July, 2007. I was standing by a creek, sipping on a can of beer, never finished it.

Standing there I looked up and saw a round rainbow, like a hula hoop, hovering in the sky. I was stunned because I had never seen a round rainbow in my life. I was in a trance. I don't remember seeing it descend or how long it was or what happened after that.

I remember next that this thing was about 5 ft. in front of me. I was afraid and excited at the same time. So I'm standing there looking at this ring, a round rainbow colored ring hovering about 3 ft. above the ground, about 5 ft. wide in diameter. The inside of the ring looked blurry compared to me looking on the outside of it, sort of watery looking. I felt electrified and afraid, like if I made the wrong move I wouldn't be here anymore. However, I managed to slowly pull my camera phone out of my pocket and took a photo. Bad news is that later on I lost it.

So I decided I needed to get out of here and I started backing away from this ring. When I got a safe distance away, I turned and ran to my car and haven't been back there since. I did manage to show the photo to a preacher friend of mine before I lost my camera phone.

Witness #2

I was near Sand Lake with my friend Jeff. We had set up a pop-up blind earlier in the day (very near a sea osprey preserve) in hopes of scoring a big buck. After midnight we walked from the back side of my grandfather's property to the area near where we had put the blind. The area of the osprey preserve is roughly the size of a football field and it is wide open wetland inside of that area. It is surrounded by trees.

As Jeff and I came in through the trees, we were about 50 feet from the open area, when we saw what we thought were two sets of headlights. I shined my spotlight at them, and Jeff got angry, because we thought they were the DNR (Department of Natural Resources), which spells trouble for two hunters in the middle of the night.

As soon as I spotlighted what we thought were ATVs the lights moved faster than anything I have ever seen, through the trees right up to twenty feet in front of us and hovered, at head level. We were both armed with full magazines, with our guns aimed at the light that was directly in front of each of us. We were standing for about 10 minutes.

We got a good look at them, but they are difficult to describe. They were the size of a volleyball and seemed to be melting neon light, pink, purple, deep blue, and green. And as I said we didn't move at all. I remember being stuck like that for what seemed like two minutes, but to Jeff it seemed more like ten minutes.

It let go of us at the same time and we fell down, and clawed dirt trying to get away, until we both had a big tree at our backs. As we were on the ground the lights had moved with us, still maintaining the 20 foot distance from us. Then, all at once, it ascended so fast you almost couldn't see it doing it, like in the blink of the eye.

This really happened. Jeff was former military with security clearance and he said that it was not man made, and had never seen anything like it. 02/09/12 23:59 CST.

Witness #3

It was so close - only about 20 foot away. It is a pulsating light, small orange/red / orange round orb light. Anyway I went outside for smokes and then saw somehow these flashes.

They are red/orange, very bright and this is not a prank. I had a chance to look and see it clearly and say, "Aha - don't FCK with me" and was able to walk back in, no problem, lock door and I know I won't be sleeping tonight at all.

Witness #4

It was 10:40 pm, the dog wanted to go out. As I opened the patio door which faces west, coming over the top of the house from the East, a bright reddish-orange orb, sphere, ball shape, about 8-9 inches in diameter with a tracer tail "stopped" 22 feet in front of me at around 7.5 feet above the ground and then blinked out.

It really scared me! I slammed the door shut and went running for my husband, I have no idea what it was, but it was too close for comfort. Needless to say, he had to let the dog back in.

Witness #5

My best friend Stephanie and I went to the movies one night and came home around twelve. We got ready for bed, went to my room on the second story of my house and opened up my window to look at the night sky and smoke a couple cigarettes as we do sometimes before we sleep. My house is in a new development, the development was built on what used to be a bunch of farms, a couple ranches still surrounding. My house is in the corner of the development. I have no neighbors to the right and only one to the left, a house slightly smaller than our own. Surrounding our house, there is very little light in the night sky, it is always a very clear.

Anyways, we're starting off towards where the river would be, enjoying our smokes and an orb shoots into the sky, perfectly...about 50 feet up and stops. We watch it hover for about a minute, the only way to describe its movements is perfect, precise, and elegant as it started to "float" or "hover" towards my neighborhood. It went behind a couple houses beyond ours and we thought "oh, well that was it" and continued with our smokes.

Still staring down my street and into the sky, a bit in awe of what we had seen, a deep eerie silence took to the surroundings, and it appears again. Only this time it was hovering, parallel to the road and moving towards my house. It was only about 20 feet in the air and slow. Of course, we started screaming the closer it got. LOUD! My sister was in the room right next to us, her bed directly on the other side of the wall where mine was. She had gone to bed. I am positive she did not have enough time to reach sleep, let alone a deep sleep. She did not wake up from our screaming. Neither did my parents, who were just down the hall and sleep with their door wide open. It was completely silent except for our terror.

As the object approached my house, let me tell you, the color was amber, yellow, and red. The orb did not glare or hurt our eyes to look at it. I shut my window, yet my blinds were still open and we stopped screaming. It hovered directly in front of my house for about 30 seconds, it seemed to be on the streetlight, but not physically harming anything. The orb then shot off, so fast, to the left of my neighbor's house and into the area behind our houses. The streetlight was out for a moment and then flickered gradually back to on.

I opened my window and we were leaning out onto the roof for just one more glimpse of this unknown object. There was a light glow coming from it and turning everything sepia colored around us. We felt so calm, and like we wanted to see "them" communicate with "them". It was a tranquil feeling, like everything would be okay. We were listening to a mix CD that had seemed to stop playing throughout this experience. We sat on my bed in awe and a song startled us a bit, from just starting to play. The lyrics are what may have alerted us - "Something tells me this isn't real, something tells me not to care." That is really how we felt, you know? Ironic, along with the fact that the CD was made that day and we just happened upon that song.

We left my window open for a while, trying to get just one more sighting, just to feel that almost comforting, calm, and tranquil feeling again. What has brought me to giving my experience to MUFON, is that I've seen more unexplained events afterwards. Orbs that turn into airplanes, some star trek/star wars looking craft hovering just 15 feet above the ground and rising slowly.

I have NEVER seen white owls until after my first experience. I know how stupid it sounds, but research I've done makes me believe owls have something more to them. I mean I've seen 7 owls take off from my roof uniformly, in a V-shape towards where I saw the first amber orb. Shortly after they were out of sight I saw a white orb, in the same spot as the amber one, fly to the same height in the sky, turn amber and almost flatten to disappear. I watched for about an hour and it did not go down my street this time, unfortunately.

There are very many witnesses to these owls, and a friend of my sister, who experienced the Phoenix Light Phenomena (amber orbs) in the nineties has had strange dreams and experiences since moving to Stanislaus County. Friends of mine, and people I don't even know, have reported seeing these amber orbs. In daytime, dusk, dawn, twilight. Everyday school kids are seeing the orbs that turn into airplanes, a new one every week. I actually saw one while trying to describe them to one of my friends, sitting in his car around the corner from my house one night. He saw this also. My sister and her boyfriend said they had seen a strange airplane at the same time I had when I got home. They had asked me before if I had said anything about whether I had seen it.

So many multiplying strange events I have had since the first. An amber orb following my sister's boyfriend home from work, just twenty feet above his truck and mimicking his speed, almost messing with him. He felt calm, and in awe. So many similar, strange and unexplainable events.

I don't know what to think anymore. I really think MUFON should research this area. I REALLY DO. Sightings and experiences are increasing as the year ends and I can't wait to see what is in store for our nation in 2007. I have never experienced a UFO or unexplainable event until 2001, and it is no shock that I found out it is the year for UFO's here. We are infested.

<u>Witness #6</u>

It was an early summer morning, about 0345. Three of my friends and I were out on a second floor balcony, talking and smoking cigarettes. There is a large tree to your right when you're standing on this balcony looking out onto the street. The neighborhood is in a condo complex near the water in Foster City, Ca.

Anyway, we were chatting away, when out of the corner of my eye I saw this orb appear from behind the tree. It arched out from behind this tree from my right to left. I was the first to notice it as it silently hovered no more than ten feet away from us at eye level. It was about 8-10 inches in diameter. The only way I can describe it is that it looked like a ball of plasma, the insides of it seemed to be moving. It was changing colors also - vivid reds and blues, pinks and orange; it was absolutely beautiful, like "living light". It didn't seem to emit light. It's hard to describe. It's like the light was internal, contained.

After a few seconds my three friends noticed it too. The conversation fell silent and all four of us just stood there dumbfounded, staring at it. I remember it seemed difficult to focus on, it's as if my eyes couldn't get a sharp look at what exactly it was. It hovered there for about 20 seconds, then it seemed to notice that we had noticed it (we all got really quiet and were staring at it at this point). The plasma ball silently floated away to our right and disappeared behind the tree where it had appeared from. We stood there shocked. Nobody said a word for a minute that seemed like an eternity.

I asked my friends "Did you guys see that?" and they said "Yeah". None of us could explain what it was. The funny thing is, none of us ever really talk about what we saw that night. I'll ask them from time to time if they remember that night, and they do. None of us can come up with an explanation of what the hell we saw.

Years later, I can't shake the feeling that there was some kind of intelligence behind it. It's hard to explain. The way it moved was very precise and deliberate. At the time, it felt like it was watching us. I think about it every day and hope to find out what it really is we saw that night.

CHAPTER FOURTEEN
HOW DO THEY GET HERE?

<u>Witness</u>

My mother, myself, and another child whom she was driving to school, were on our way to school/work around 9:30-10:00am. About 5 miles from my mothers work, we noticed rainbows, very short - about the size of the sun, on either side of the sun. We observed this for about 3 minutes. When we began to turn onto the highway, we saw a small black swirl appear in the center of the right rainbow. The hole stayed open for a few minutes as we parked.

We stepped out and continued to observe the rainbows and black hole. After a few minutes we noticed something beginning to peek from the hole. It was silver and looked as if it was trying to roll uphill, under weak power before exiting the rainbow/hole. The silver ball object soon exited the black hole which closed behind this object. The object hovered at the exit point for about a minute. I could see small red points on its outside. It then, very slowly, began to descend towards the ground, finally disappearing behind the tree line.

We waited and watched for about 10-15 minutes afterwards and never saw the object again. It appeared to be about 20 miles away. I would estimate its size to be roughly 60 feet in circumference. We were all mesmerized by the sighting, but suffered no adverse effects.

PART THREE

WHAT AND WHY?

CHAPTER ONE
WHAT?

Based upon the vast amount of information presented in the first two parts of this book, what can we conclude about the orange orb phenomenon? There are likely as many different conclusions as there are readers. As I quickly learned when I was a young trial lawyer, from the same body of facts, each member of the jury will draw different conclusions. However, if enough of them have conclusions that are somewhat similar, they may have sufficient agreement to form a consensus.

What I offer below are the conclusions I have drawn from what I have learned about the orange orbs. Knowing full-well that others may draw conclusions different from mine, perhaps there may be enough similarity to form a general consensus among us on the truth and reality of the worldwide invasion of the orange orbs.

******

1) There are orbs flying above and landing on Earth.

2) These orbs appear not to be of Earthly origins.

3) They are piloted/controlled by intelligent visitors from a planet other than Earth.

4) They travel to Earth by use of "wormholes" (holes created through the space/time fabric to reduce the gap between very distant points in the universe to a very short distance – as theorized by Albert Einstein).

5) The orbs fly throughout the entire airspace of Earth.

6) The number of orbs visiting the Earth is increasing around the world at a very rapid rate.

7) The orb flights and landings are well-planned.

8) The planning includes specific underwater bases that are strategically placed around the world.

9) The orbs from each base are assigned to visit, on a regular basis, various geographic areas of the world.

10) Those piloting/controlling the orbs carry-out maneuvers and flight plans which are designed to attract attention to themselves and be seen by as many human beings on earth as possible.

11) They do this by a) cloaking themselves in very bright and pleasant colors, b) choosing to fly over the most heavily-populated areas of the world at the times of the evening when the sky is dark so their bright coloring will attract attention and when most human beings are still awake (8:30 – 10:30 PM), c) flying low, silently (to differentiate themselves from Earth vehicles) and very slowly, d) engaging in flight patterns that will attract attention (V formations, trail formations, mimicking constellations, "dancing" in the sky, flashing lights of various colors, etc.), e) making specific appearances on the days of the year when nearly everyone in the U.S. is very likely looking at the sky—just after the 4th of July and the New Year's Eve fireworks (the two days of the year with the most orange orb sighting, by far).

12) The orbs emit beams of varying colors which are used, at times, to convey objects and orb passengers to the Earth's surface and to extract items from the surface to the orb.

13) The vehicles within the colored cloaks are of a solid material.

14) The vehicles within the colored cloaks are of a variety of construction, size, and shape.

15) There are very large "mother ships" among the vehicles that accompany the orbs that are very dark and triangular in shape.

16) The orbs attach themselves at various times and on various locations to the "mother ship's" exterior.

17) The vast majority of the sighted orbs are orange/red in color but there are also red, white, green, blue, and "rainbow" orbs sighted.

113

18) The orbs are of various sizes from hundreds of feet in diameter to baseball size.

19) The smaller orbs appear to be a type of drone which come very close to human beings on the Earth and seem to be observing and studying them.

20) Earth animals—dogs in particular—have very strong reactions to the appearance and/or sounds of the orbs.

21) Orbs will, on occasion, react to and/or mimic human actions.

22) Orbs have been seen to repeat the pattern of flashlight blinks produced by a human being and they display an avoidance of laser lights focused on them, frequently by instantly "disappearing" when this occurs.

23) The orbs seem to be attracted to the color/heat of a lighted cigarette. There are far too many cases of lighted cigarettes and corresponding orb appearances to be explained by mere coincidence.

24) Orange orbs are often seen discharging a type of glowing/glittering orange material as they cross the night sky. None of this material has, as yet, been detected on the ground.

25) Orange orbs have often been observed entering and exiting bodies of water. The orange light from the orb has been seen reflecting on the water's surface and water ripples have been observed as the craft enters the water. The orb has been seen shutting down its colored cloak before entering the water, revealing a solid craft underneath.

26) The orbs tend to enter the water at a very slow rate but emerge skyward from the water at extreme velocities.

27) The orbs appear to "split" into numerous smaller orbs then "merge" to reform the original larger orb.

28) Given that the vehicles are made of solid materials, it is unlikely that the vehicles, themselves, actually "split" or "merge." It is much more likely that an orb's colored cloak can expand to any given size and can contain any number of vehicles within a single cloak. When "splitting," these vehicles emerge from the large single cloak and appear outside this "mother" cloak, covered with their own individual cloaks. When "merging," these vehicles return to within the "mother" cloak, seeming to become a part of the orb to the observer.

29) The crew of the orbs cloak their vehicles by using a mechanism at the bottom of the craft which generates the cloaking substance, which emerges from this mechanism and which moves upward around the vehicle until the entire vehicle is covered in the spherical cloak.

30) The crew can choose their cloak coloring and can, instantly, turn it off or on.

31) The government is well-aware of the orange orb phenomenon and military aircraft frequently appear in the near vicinity of orb appearances and often give chase to the orbs to no avail. The orbs quickly "blink out" when military vehicles appear or move away at speeds far beyond the capability of Earth-made aircraft.

32) The government makes no admissions of its awareness of the orange orb phenomenon and, instead, offers ludicrous explanations to account for their appearances.

33) There have been no reports filed that indicate any violent intent by the visitors.

34) There have been reports of the visitors "capturing" human beings and causing them to undergo procedures described as "disgusting" and/or "very unpleasant." But, other than resultant "bad memories" and psychological trauma, there have been no reports of actual human physical injuries or death at the hands of the visitors.

35) There have been numerous sightings of the passengers and crew of the orbs both on-the-ground sightings and through the "windshields" of their vehicles.

36) The descriptions of these visitors are very consistent.

37) There appear to be two types of visitors with either very large or rather small bodies.

38) The interaction between the large and small visitors seems to connote a type of parent/child relationship.

39) The visitors are capable of projecting real-looking entities that appear to be human and are used in interactions with human beings. These human-looking entities do not appear to speak, nor do the actual visitors, large or small.

40) The visitors appear to communicate with human beings without speaking.

41) The tall visitors are uniformly described as being human-shaped, very tall (7 to 8

feet in height), very thin and muscular with exceptionally long arms and fingers and legs, entirely covered with a shiny, skin-clinging, black material that has a luminescence, crown-shaped heads, sometimes wearing goggles, eyes that are very human-like but larger, a type of covering on their heads that appears to be either a type of hair or flaps of skin, capable of running with extreme speed, possess tremendous strength, having very small noses and a slit for a mouth and no apparent ears.

42) The small visitors also have human-like bodies that are thin. Mouths and noses are similar to the tall visitors as is the lack of ears. Their heads look larger in proportion to their bodies than do the tall visitors.

43) The visitors are often quite indifferent to being seen by human beings.

44) When faced with human aggression, the visitors choose retreat as a response. Given the visitors complete avoidance of the Washington D.C. airspace because of the plethora of anti-aircraft weapons/aircraft in position there, it appears that they have a general intent of avoiding confrontation with human beings.

45) The visitors can cloak themselves with a type of invisibility when they so choose, making it difficult, but not impossible, to be seen.

46) The visitors can employ a type of anti-gravity mechanism to suspend their bodies at various elevations above the ground and are able to walk at these elevated heights using very awkward body movements to do so.

47) When running, the visitors exhibit an unusually upright and straight posture of their upper bodics as opposed to the forward leaning stance of human beings.

48) Human beings have a wide variety of emotional reactions to orb/visitor sightings, from awe and a sense of well-being to fascination, to puzzlement, to fear and sometimes to outright terror.

49) Some witnesses look forward to future sightings/interactions with the visitors while others wish to never see them again.

50) Our cosmic visitors not only emerge from their space vehicles to walk about the Earth, but also have entered our lawns, porches, and the interior of our homes.

51) The number of orb/craft landings and observations of the cosmic visitors on the ground has dramatically increased over the past two years.

CHAPTER TWO
WHY?

Perhaps the most fascinating aspect of the orange orb phenomenon is the question "Why are they here?". While we can present the facts about orange orbs and work on a consensus about what we know about the phenomenon, the question of "why?" is an entirely different matter.

The "why?" question requires us to attempt to enter the minds of our cosmic visitors and discern their thinking in regard to their choosing Earth as a planet they wish to visit and their intentions for coming here.

This process is entirely speculative but worthy of pursuing since we are not working in a vacuum in this process. We know quite a bit about the visitors' behaviors toward Earth and human beings, so the present task is to attempt to derive their thinking by analyzing what they are doing. Through this process, we can, at least, arrive at some plausible suggestions as to why they are here.

With this in mind, I have set forth below what I believe is plausible speculation, based upon the information we have, as to why our visitors are here on Earth.

******

In my opinion, the key factor in developing a plausible explanation as to why we are being visited by our cosmic visitors is that they clearly want to be seen by as many human beings as possible.

For most of the documented history of UFOs, the non-Earth craft have been quite elusive with rare sightings of them and even rarer photographic evidence. With the advent of the orange orbs, however, all of this has changed. The orange orbs not only are not trying to be elusive but are doing just the opposite—they are going out of their way to be seen by as many human beings as possible. As for photographic evidence, there is a flood of it with thousands of excellent photos and videos of the orbs and some of the visitors themselves.

We must ask ourselves, why this dramatic change in behavior? They are not surreptitiously trying to spy on us and secretly collecting research samples as they were probably doing in the past. They are, instead, clearly shouting in their own way, "Here we are … look at us!"

Also, as an integral part of what they are doing, they have chosen to cloak their vehicles in beautiful, pastel, muted, inner-lighted colors that often precisely mimic a setting sun. And they do things in the sky that are entertaining and very non-threatening. And they choose to appear as presenting themselves as posing no danger to us.

Why? And why have they become so numerous so as to constitute a worldwide invasion?

If another human being were behaving in a similar way, such as this, it is likely that we would conclude that this person wants us to like him and wants us to get used to seeing him and not be afraid of him. So why would we not conclude the same about our cosmic visitors?

In my opinion, this is exactly why our visitors are behaving the way that they are.

So, if we use this as our working hypothesis, we must then ask ourselves why would they want us to like them and get used to them and not be afraid of them?

I suggest the following:

These are obviously very intelligent cosmic beings with a far broader knowledge of many things that are beyond human comprehension. Also, I believe it can be reasonably assumed that Earth is not the first planet they have visited and that they are quite experienced in dealing with planets with sentient populations who are as primitive as the human beings on Earth.

With their broad experience, our visitors must be quite aware that an inappropriate intrusion by an advanced race of beings into a primitive civilization can be disastrous to that population. Such a cultural shock can undermine the fundamental traditions, beliefs, and self-perceptions of the people's place in the world. If a people lose these things, they cease to exist as a people and all social and religious bonds are broken, causing them to become unattached and lost strangers.

We are rife with examples of this in our Earth history: the Spanish invasion of the Aztecs, Incas, and Mayans, the European invasion of the Native Americans, the intrusion of the Europeans into the Aborigines of Australia, and Margaret Meade's intrusive research into primitive island tribes of the South Pacific.

For eons, we on Earth have undoubtedly been visited by cosmic travelers. When human beings were quite primitive, localized intrusions could probably be safely managed. Once the Earth's human populations reached a point of world communications and awareness, however, intrusions became much more precarious because of the likelihood that local news can now become world news very quickly.

This is the reason, I believe, why cosmic visitors have remained as invisible as possible for the past century. Something has dramatically changed in their thinking quite recently, however—within the last decade. It is clear that the visitors have made the decision to come out of hiding and begin the process of slowly introducing themselves to the sentient population on Earth. Who can be sure just what has occurred on Earth to bring the visitors out of hiding? But, it is clear they are—and as gently as possible—they seem to be making a distinct effort to try to avoid as much cultural disintegration as possible.

In many classic sci-fi space movies, the plot was that cosmic visitors would seek out world leadership to announce their existence. It appears they got it wrong. What is currently happening with our orb visitors is that they are making their introductions directly to the people and avoiding the government at all costs—flying away when they show up.

So, if this is what is going on, what is it leading to?

My opinion is that the orb visitors will continue to intensify their worldwide appearances until they reach a critical mass when the awareness of their presence is general knowledge throughout the world, so that even the governments of the world must drop their pretense that these orb visitors do not exist.

Once this point is reached and the orb visitors believe that they can advance to the next stage of their introduction process, they will initiate it. I believe it will begin with greatly increased landings and increased contact with the human race, done in such a way as to not cause general panic or fear. It will be, in essence, a public relations campaign, which they have, no doubt, carried out countless times with the races of other planets, when the visitors believe they can be safely integrated. No doubt, they will emphasize their peaceful intentions and the benefits they can bestow upon the human race.

One sure result will ensue—all humans will drop what they are doing, be it wealth building, divorcing, making love or war, whatever, and will be overwhelmed by a new perspective where such things seem to be too small for further consideration, at least for the moment. If our orb visitors declare that war must end everywhere on Earth, it will, because they, like a father telling his kids to "break it up," have the power to make sure they do.

And, perhaps, I am an unknowing participant in their plans, helping to alleviate the fears of their growing presence and their soon-coming mass touchdowns.

I am OK with this, as long as they don't ask for a share of my book royalties.